AF361826

Forestry Applications of Unmanned Aerial Vehicles (UAVs) 2019

Forestry Applications of Unmanned Aerial Vehicles (UAVs) 2019

Editor

Alessandro Matese

MDPI • Basel • Beijing • Wuhan • Barcelona • Belgrade • Manchester • Tokyo • Cluj • Tianjin

Editor
Alessandro Matese
National Research Council (CNR-IBE)
Italy

Editorial Office
MDPI
St. Alban-Anlage 66
4052 Basel, Switzerland

This is a reprint of articles from the Special Issue published online in the open access journal *Forests* (ISSN 1999-4907) (available at: https://www.mdpi.com/journal/forests/special issues/ UAVs Applications).

For citation purposes, cite each article independently as indicated on the article page online and as indicated below:

LastName, A.A.; LastName, B.B.; LastName, C.C. Article Title. *Journal Name* **Year**, *Article Number*, Page Range.

ISBN 978-3-03936-754-2 (Hbk)
ISBN 978-3-03936-755-9 (PDF)

Cover image courtesy of Alessandro Matese.

Contents

About the Editor

Alessandro Matese, Dr., is a Researcher at National Research Council—Institute of Bioeconomy (CNR-IBE). He received his M.S. degree in Natural Sciences at University of Florence (Italy), Department of Earth Sciences, and his Ph.D. degree in Agriculture, Forest and Food Science at University of Turin (Italy), Doctoral School of Sciences and Innovative Technologies. His research is focused on remote sensing of agroecosystems, precision agriculture and forestry, unmanned aerial vehicles, multi-hyperspectral and thermal imaging, crop modelling, data fusion, machine learning, computer vision, and geostatistics. Dr. Matese has authored and co-authored more than 50 peer-reviewed international journal articles. He has been a keynote speaker and invited speaker at numerous international and national conferences and meetings. Dr. Matese is Editor of the MDPI journals Forests and Remote Sensing and Associate Editor of Frontiers in Plant Science, Technical Advances in Plant Science section.

Editorial

Editorial for the Special Issue "Forestry Applications of Unmanned Aerial Vehicles (UAVs)"

Alessandro Matese

Institute of BioEconomy (IBE), National Research Council (CNR), Via Caproni 8, 50145 Florence, Italy; alessandro.matese@cnr.it; Tel.: +39-055-3033711

Received: 2 April 2020; Accepted: 3 April 2020; Published: 5 April 2020

Abstract: Unmanned aerial vehicles (UAVs) are new platforms that have been increasingly used in the last few years for forestry applications that benefit from the added value of flexibility, low cost, reliability, autonomy, and capability of timely provision of high-resolution data. This special issue (SI) collects nine papers reporting research on different forestry applications using UAV imagery. The special issue covers seven Red-Green-Blue (RGB) sensor papers, three papers on multispectral imagery, and one further paper on hyperspectral data acquisition system. Several data processing and machine learning methods are presented. The special issue provides an overview regarding potential applications to provide forestry characteristics in a timely, cost-efficient way. With the fast development of sensors technology and image processing algorithms, the forestry potential applications will growing fast, but future work should consider the consistency and repeatability of these novel techniques.

Keywords: unmanned aerial vehicles (UAV); precision forestry; forestry applications; image processing; machine learning; RGB imagery

---

1. Introduction

Unmanned aerial vehicles (UAVs) are new platforms that have been increasingly used in the last few years for forestry applications that benefit from the added value of flexibility, low cost, reliability, autonomy, and capability of timely provision of high-resolution data. The main adopted image-based technologies are RGB, multispectral, and thermal infrared. LiDAR sensors are becoming commonly-used to improve the estimation of relevant plant traits. In comparison with other permanent ecosystems, forests are particularly affected by climatic changes due to the longevity of the trees, and the primary objective is the conservation and protection of forests. Nevertheless, forestry and agriculture both involve the cultivation of renewable raw materials—the difference is that forestry is less tied to economic aspects and this reflects the delay in using new monitoring technologies. The use of UAV in precision forestry has exponentially increased in the last years as demonstrated by a large number of papers published between 2018 and 2019, and more than 400 references are found searching for "UAV" + "forest" and considering articles, conference proceedings, and books.

The main forestry applications aim to inventory resources, map diseases, species classification, fire monitoring, and spatial gaps estimation. This Special Issue focused on new technologies (UAV and sensors) and innovative data elaboration methodologies (object recognition and machine vision) for forestry applications.

2. Overview of Contributions

VOSviewer software version 1.6.14 (Centre for Science and Technology Studies, Leiden University, The Netherlands) was used for a simple bibliometric map, shown in Figure 1. The software was developed for creating, visualizing, and exploring papers' bibliometrics. Term map offers overviews for identifying the structure of a topic and represents research topics where strongly related terms are located close to each other and the weaker the relationship is between terms, the greater the distance is between them. Terms were extracted in titles and abstracts of papers and represented in the map as circles.

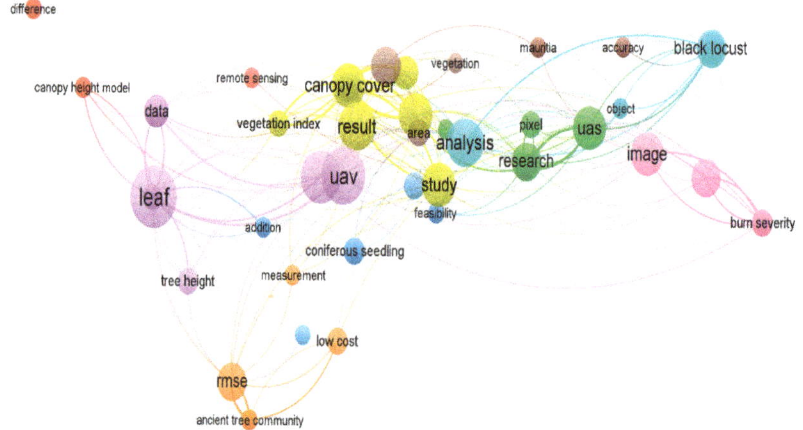

Figure 1. Term clustering map based on the SI publications. Different colors represent the terms belonging to different clusters. The size of the term is based on the number of occurrences. The connecting lines indicate the strongest co-occurrence links between terms.

Seven papers in this special issue reported results from UAV platform using RGB cameras [1–7], three paper used multispectral cameras [6,8,9], while one paper used a hyperspectral camera [6].

The aims of the papers were to test the feasibility of using UAVs to rapidly identify coniferous seedlings in replanted forest-harvest using an efficient sampling-based approach, consumer-grade cameras, and straightforward image handling, such as in [1]. The tree characteristics monitored were tree height, crown width, prediction of diameter at breast height (DBH), and tree age with low cost, high efficiency, and high precision in [2]. The development of a dataset called MauFlex related to Mauritia flexuosa palm, also known as "aguaje", was a study aimed at its conservation as it is a species poorly monitored because of the difficult access to these swamps. Moreover, a segmentation and measurement method for areas covered in Mauritia flexuosa palms using high-resolution aerial images acquired by UAVs was performed in this research [3]. One study evaluated the feasibility of adopting the low-cost, flexible, high-resolution, sensor-capable UAV platform for collecting reference data to use in thematic map accuracy assessments for complex environments [4]. Another research study focused on seed germination, stump shoot resprout, and spreading by root suckering of black locust in ten short rotation coppices [5]. The evaluation of density and canopy cover of western juniper in a treated (juniper removed) and an untreated watershed and an assesment of the effectiveness of using low altitude UAV-based imagery to measure juniper-sapling population density and canopy cover were also studied [6]. The investigation of UAV-based photogrammetric point clouds and hyperspectral imagery for characterizing seedling stands in leaf-off and leaf-on conditions were researched in [7]. The study of multispectral UAV images that can be used to classify burn severity, including the burned surface class, were considered in [8]. Lastly, the estimation of Chestnut pruning biomass through differences in the volume of canopy trees and an evaluation of the performance of an unsupervised segmentation methodology as a feasible tool for the analysis of large areas were considered in [9].

Various data processing methods were used in this Special Issue.

Image analysis workflow, where a three-step, object-based process consisting of image segmentation, automated classification using a classification and regression tree (CART) machine-learning algorithm, and the merging of adjacent image objects classified as 'seedlings' into single seedling objects, was conducted in [1]. The Structure-from- Motion/Multi-view Stereo (SfM-MVS) method was used to detect the feature points of the image, and the scale-invariant feature transform (SIFT) algorithm was used to detect feature points and generate feature vectors. Then, matching was carried out according to the feature vectors, and the RANSAC (random sample consensus) algorithm was used to delete the connection of the conflicting geometric features of corresponding feature points in [2]. Convolutional Neural Network (CNN) based on the Deeplab v3+ architecture was completed. Images were acquired under different environment and light conditions using three different RGB cameras in [3]. Pixel-based and object-based classification reference data methods were used in [4]. Object-based image analysis (OBIA) and convolutional neural network (CNN) were used in [5]. Vegetation indices and support vector machine classification were used in [6]. Watershed segmentation method to delineate the tree canopy boundary at an individual tree level, and optimal bands for calculating vegetation indices were determined. Species classification using the random forest method was used in [7]. Maximum likelihood (MLH), spectral angle mapper (SAM), and thresholding of a normalized difference vegetation index (NDVI) were used as classifiers in [8]. Supervised and unsupervised crown segmentation with a double filtering process based on Canopy Height Model (CHM) and vegetation index threshold were used in [9].

Conflicts of Interest: The authors declare no conflict of interest.

References

1. Feduck, C.; McDermid, G.; Castilla, G. Detection of Coniferous Seedlings in UAV Imagery. *Forests* **2018**, *9*, 432. [CrossRef]
2. Qiu, Z.; Feng, Z.; Wang, M.; Li, Z.; Lu, C. Application of UAV Photogrammetric System for Monitoring Ancient Tree Communities in Beijing. *Forests* **2018**, *9*, 735. [CrossRef]
3. Morales, G.; Kemper, G.; Sevillano, G.; Arteaga, D.; Ortega, I.; Telles, J. Automatic Segmentation of Mauritia flexuosa in Unmanned Aerial Vehicle (UAV) Imagery Using Deep Learning. *Forests* **2018**, *9*, 736. [CrossRef]
4. Fraser, B.; Congalton, R. Evaluating the Effectiveness of Unmanned Aerial Systems (UAS) for Collecting Thematic Map Accuracy Assessment Reference Data in New England Forests. *Forests* **2019**, *10*, 24. [CrossRef]
5. Carl, C.; Lehmann, J.; Landgraf, D.; Pretzsch, H. *Robinia pseudoacacia* L. in Short Rotation Coppice: Seed and Stump Shoot Reproduction as well as UAS-based Spreading Analysis. *Forests* **2019**, *10*, 235. [CrossRef]
6. Durfee, N.; Ochoa, C.; Mata-Gonzalez, R. The Use of Low-Altitude UAV Imagery to Assess Western Juniper Density and Canopy Cover in Treated and Untreated Stands. *Forests* **2019**, *10*, 296. [CrossRef]
7. Imangholiloo, M.; Saarinen, N.; Markelin, L.; Rosnell, T.; Näsi, R.; Hakala, T.; Honkavaara, E.; Holopainen, M.; Hyyppä, J.; Vastaranta, M. Characterizing Seedling Stands Using Leaf-Off and Leaf-On Photogrammetric Point Clouds and Hyperspectral Imagery Acquired from Unmanned Aerial Vehicle. *Forests* **2019**, *10*, 415. [CrossRef]
8. Shin, J.; Seo, W.; Kim, T.; Park, J.; Woo, C. Using UAV Multispectral Images for Classification of Forest Burn Severity—A Case Study of the 2019 Gangneung Forest Fire. *Forests* **2019**, *10*, 1025. [CrossRef]
9. Di Gennaro, S.; Nati, C.; Dainelli, R.; Pastonchi, L.; Berton, A.; Toscano, P.; Matese, A. An Automatic UAV Based Segmentation Approach for Pruning Biomass Estimation in Irregularly Spaced Chestnut Orchards. *Forests* **2020**, *11*, 308. [CrossRef]

Article

An Automatic UAV Based Segmentation Approach for Pruning Biomass Estimation in Irregularly Spaced Chestnut Orchards

Salvatore Filippo Di Gennaro [1], Carla Nati [1,*], Riccardo Dainelli [1], Laura Pastonchi [1], Andrea Berton [2], Piero Toscano [1] and Alessandro Matese [1]

[1] Institute of BioEconomy (IBE), National Research Council (CNR), Via Caproni 8, 50145 Florence, Italy; salvatorefilippo.digennaro@cnr.it (S.F.D.G.); riccardo.dainelli@ibe.cnr.it (R.D.); laura.pastonchi@ibe.cnr.it (L.P.); piero.toscano@cnr.it (P.T.); alessandro.matese@cnr.it (A.M.)

[2] Institute of Clinical Physiology (IFC), National Research Council (CNR), Via Moruzzi 1, 56124 Pisa, Italy; bertonandrea@ifc.cnr.it

* Correspondence: carla.nati@ibe.cnr.it; Tel.: +39-055-5225640

Received: 23 December 2019; Accepted: 10 March 2020; Published: 12 March 2020

Abstract: The agricultural and forestry sector is constantly evolving, also through the increased use of precision technologies including Remote Sensing (RS). Remotely biomass estimation (WaSfM) in wood production forests is already debated in the literature, but there is a lack of knowledge in quantifying pruning residues from canopy management. The aim of the present study was to verify the reliability of RS techniques for the estimation of pruning biomass through differences in the volume of canopy trees and to evaluate the performance of an unsupervised segmentation methodology as a feasible tool for the analysis of large areas. Remote sensed data were acquired on four uneven-aged and irregularly spaced chestnut orchards in Central Italy by an Unmanned Aerial Vehicle (UAV) equipped with a multispectral camera. Chestnut geometric features were extracted using both supervised and unsupervised crown segmentation and then applying a double filtering process based on Canopy Height Model (CHM) and vegetation index threshold. The results show that UAV monitoring provides good performance in detecting biomass reduction after pruning, despite some differences between the trees' geometric features. The proposed unsupervised methodology for tree detection and vegetation cover evaluation purposes showed good performance, with a low undetected tree percentage value (1.7%). Comparing crown projected volume reduction extracted by means of supervised and unsupervised approach, R^2 ranged from 0.76 to 0.95 among all the sites. Finally, the validation step was assessed by evaluating correlations between measured and estimated pruning wood biomass (Wpw) for single and grouped sites ($0.53 < R^2 < 0.83$). The method described in this work could provide effective strategic support for chestnut orchard management in line with a precision agriculture approach. In the context of the Circular Economy, a fast and cost-effective tool able to estimate the amounts of wastes available as by-products such as chestnut pruning residues can be included in an alternative and virtuous supply chain.

Keywords: unmanned aerial vehicles; precision agriculture; biomass evaluation; image processing; *Castanea sativa*

1. Introduction

Remote Sensing (RS) is one of the technologies that has been currently most employed in the forestry sector for monitoring, inventorying, and mapping purposes. RS techniques with the aim to obtain information on large areas can be conducted at different levels of precision, according to the different goals to be achieved. The choice of the RS platform to be employed, and consequently the sensors installed and operating on-board that specific platform will depend on the processes under investigation and the level of detail required for a particular analysis [1].

RS platform as satellite systems, aircraft platforms and unmanned aerial vehicles (UAVs) have features that differ in terms of spatial resolution, surface covered, temporal resolution, operational procedures, and costs. Satellite solutions remain a fundamental tool for long-term and extensive monitoring and surveillance forestry activities against fire events [2], pests attack [3], illegal logging [4] and more generally, to assess the health and structure of forests' cover [5]. Aircraft platforms provide a better image resolution, returning a higher level of detail compared to satellite, against a higher effort in flight planning and relevant operational costs [6]. UAVs are flexible small platforms characterized by low operational costs, high spatial and temporal resolution [7] but suitable to cover only limited areas. Comparisons among different platforms have been made both in the agricultural [8] and in the forestry field [9].

The use of UAV in precision forestry has exponentially increased in recent years, as demonstrated by the large number of papers published between 2018 and 2019; more than 400 references were found when searching for "UAV" + "forest" and considering articles, conference proceedings and books [10].

Authors have dealt with several research topics involving applications in forest monitoring, inventorying, and mapping both with multirotor and fixed-wing unmanned platforms equipped with a wide series of optical technology sensors [11–18]. These studies took into account forestry UAV applications mainly within two forest types: the first one included planted, pure and even-aged forests [19–24] and the second one included natural, mixed and uneven-aged forests where the spatial variability of vegetation was very high [25–29].

Within natural, mixed and uneven-aged forests research, UAVs have been employed most commonly for (i) estimation of dendrometric parameters such as dominant height, stem number, crown area, volume and above-ground biomass (Wa) using RGB (Red–Green–Blue bands camera) [30–34], multispectral near red green (NRG) [35,36] and laser scanning [37,38] sensors. This is the top research topic because reliable information on the status and trends of forest resources is the basis for the decision-making process for forest management and planning [39]; (ii) tree species classification and invasive plants detection for forest inventories and monitoring of biodiversity using RGB [40,41], multispectral [42,43], hyperspectral [29,44] and laser scanning [45] sensors; (iii) flight plan ad RGB sensor settings to improve imagery products accuracy [26,46–49] (iv) forest health monitoring and diseases mapping using different sensors (RGB [50], multispectral [51], hyperspectral [52], thermal [15]) to provide data for supporting intervention decisions in the management of forests; (v) recovery monitoring after fire events or conservation interventions through UAV equipped with RGB [53] and multispectral [2,17] cameras.

By providing key forest structural attributes such as tree crown centers and boundaries, UAV imagery tree segmentation is used for stem counting [32,54], extrapolation of further dendrometric parameters (i.e., Wa) [55–59], species recognition [42], and pathogens detection and mapping [60].

Regarding Wa estimation, there are two main strategies adopted for Digital Aerial Photogrammetry (DAP) and Airborne Laser Scanning (ALS)-based analysis in forestry inventories: (i) the Area-Based Approach (ABA), a distribution-based technique which provides data at stand level using predictive models developed with co-located ground plot measurements and RS data that are then applied to the entire area of interest to generate estimates of specific forest attributes [61]; and (ii) Individual Tree Crown segmentation (ITC) delineation, in which individual tree crowns, heights and positions are the basic units of assessment and where specific algorithms are used to identify the location and size of individual trees from raster images or high-density point clouds [62]. Previous research papers dealt with biomass estimation both at the stand and at tree level. Biomass at stand level is evaluated by comparing the effects of flight settings, sensor type and resolution in tropical woodlands [55], the influence of plot size in dry tropical forests [58] or by taking into account different mangrove species in South China wetlands [59]. Concerning tree-level biomass estimation, Guerra-Hernandez et al. [57] and Guerra-Hernandez et al. [56] used, respectively, UAV-DAP point clouds in open Mediterranean forest of coniferous *Pinus pinea* (Central of Portugal) and DAP and ALS data in evergreen *Eucalyptus spp.* plantation (North of Portugal). The latter two studies are important references for modeling SfM individual tree diameters and SfM-derived individual tree biomass (Wa_{SfM}) and volume (V_{SfM}) from the canopy height model (CHM) in Mediterranean forest plantation.

Segmentation of individual tree crowns is difficult, particularly in broadleaf, mixed, or multi-layered forests. This is generally due to an inability to determine the appropriate kernel size to simultaneously minimize omission and commission error with respect to tree stem identification [63]. In the literature, several unsupervised segmentation approaches have been proposed: the most widely used is the watershed segmentation algorithm [20,23,34,64–67] and its variants [16,54,68]. Other techniques are multiresolution segmentation algorithm [27,69], large-scale mean-shift algorithm [35], semantic-level segmentation using a Convolutional Neural Network (CNN) [70] and more complex approaches with two or more integrated algorithms [37,63,71]. Some authors associated the above-mentioned unsupervised approaches to manually drawn individual tree crown polygons from on-screen interpretation to compare and validate results or provide a reference for the accuracy assessment of an automatic procedure [72–74].

Among the papers that adopted both manual and unsupervised tree segmentation, only a few research works included ground data collection [42,72,75–77] with a tree sample size ranging from 109 to 2069 trees. None of those presented wood biomass in-field data. For natural, mixed and uneven-aged forest, Mayr et al. [75] gathered tree height in dry savannah and used an implementation of the watershed segmentation algorithm provided by System for Automated Geoscientific Analyses-Geographic Information System (SAGA-GIS) while Franklin and Ahmed [42] utilized the multi-resolution segmentation procedure with the ENVI software system and they collected tree height and crown dimensions in a mixed maple, aspen, and birch forest. Concerning planted, pure and even-aged forests, Ganz et al. [72] used a multiresolution segmentation algorithm and measured tree height within stands of Norway spruce and common beech while Apostol et al. [77] utilized the watershed algorithm and collected tree height and stem diameter in an even-aged Douglas fir stand. By taking tree height as ground-truth data in a chestnut plantation, Marques et al. [76] segmented trees by combining a vegetation-index based algorithm with the Otsu method.

Chestnut (*Castanea sativa Mill.*) orchards are a type of multifunctional tree cultivation used worldwide that represent a relevant income for rural populations. In Italy, sweet chestnut groves cover 147,568 hectares (ha) of the whole Italian forested territory [78]. Only a few research papers used UAV in chestnut plantations and dealt with phytosanitary problem detection and monitoring of tree health [76,79,80], automatic classification and segmentation of chestnut fruits through Convolutional Neural Networks (CNNs) [81], and insects damage rate detection and pest control methods [82]. However, there is no research available that tried to estimate the amounts of residues coming from tree tending by using UAV techniques and comparing their information with ground truth. In the present study, the authors applied RS techniques (UAV) to collect data on uneven-aged and irregularly spaced chestnut (*Castanea sativa Mill.*) orchards. The aim of the present study was to verify the reliability of RS techniques for the estimation of pruning wood biomass (Wpw) through differences in the volume of canopy trees calculated with supervised extraction and to evaluate the performance of an unsupervised segmentation methodology as a feasible tool for large-area analysis. In the context of the Circular Economy, a fast and cost-effective tool able to estimate the amounts of residues available as by-products, such as chestnut pruning material, can be included in an alternative and virtuous supply chain.

2. Materials and Methods

2.1. Experimental Sites

The study took place within the Amiata mountain region (Tuscany, Italy) between 2017 and 2018. Four sites located into three different chestnut orchards were selected as representative of this area in terms of variety and management practices (Figure 1).

Figure 1. Experimental sites map.

The experimental sites' characteristics are shown in Table 1. The chestnut orchards under study were uneven-aged and irregularly spaced (Figure 1), mainly due to replacements of dead trees that occurred over time. For this reason, they were equated to irregular forests.

Table 1. Experimental sites' description.

ID	Site A	Site B	Site C	Site D
Location	42°53'18.22" N 11°33'41.57" E	42°53'17.19" N 11°33'41.75" E	42°52'11.71" N 11°30'28.55" E	42°52'18.59" N 11°30'29.65" E
Altitude (m ASL)	960	1085	780	755
Surface (ha)	0.55	0.32	0.36	0.32
Chestnut variety	Cecio	Cecio	Bastarda Rossa	Bastarda Rossa
Density (trees ha^{-1})	72.57	114.60	110.10	111.73
Canopy cover (%)	82.50	86.47	90.19	87.80

2.2. Pruning Wood Biomass Ground Measurement

At the beginning of March 2017, 30 chestnut trees per plot (A, B, C, D) were selected and the diameter at breast height (DBH) was callipered. The choice of the sample trees was made by identifying plants representative of each site in terms of size (DBH). Sample trees were georeferenced at high resolution (0.02 m) with a differential GPS (Leica GS09 GNSS, Leica Geosystems AG). In February 2018, the previously selected trees were pruned, their branches severed and grouped in two sets: "wings" (pieces below 4 cm in diameter) and "wood" (pieces over 4 cm in diameter). The first group had no commercial use while the second could follow two different destinations: sold as firewood after being seasoned in the field or sold to the industry for tannins extraction. These two raw materials were separately loaded on a tractor equipped with a bucket or a pitchfork and weighed by means of portable scales (model WWSD6T, Nonis s.r.l., Biella, Italy). Every five weighing the scales' accuracy was checked by weighing the tractor unloaded. From every site, Wpw samples were collected and weighed fresh, then oven-dried according to the standard UNI EN ISO 18134-2:2017 to measure their moisture content. The following analyses and comparisons were made on a dry matter basis, avoiding uncontrollable variability due to wood samples size, initial moisture conditions or seasoning.

2.3. UAV Platform and Data Processing

Remote sensed data were acquired to characterize the intra-plot variability in terms of plant vigor and Wpw. Two flight campaigns were performed on 2 August 2017 and on 25 July 2018 at the same phenological stage using a modified multi-rotor Mikrokopter (HiSystems GmbH, Moomerland, Germany) described in Matese and Di Gennaro [83] equipped with a multispectral camera Tetracam ADC Snap (Tetracam Inc., Chatsworth, CA, USA). The second flight was performed immediately after the canopy pruning management to enable a comparison between ground truth and UAV results. Multispectral image acquisition was planned flying at 60 m above ground level at midday, yielding a ground resolution of 0.05 m pixel^{-1} and a 70% overlap in both directions. The images were recorded in clear sky conditions. The radiometric calibration processes were realized by acquiring, during the flight, images from three OptoPolymer (OptoPolymer-Werner Sanftenberg, Munich, Germany) reference panels, with 95%, 50%, and 5% reflectance, respectively.

The data processing workflow is described in Figure 2. Multispectral or NRG images with three broad bands (Near-infrared–Red–Green bands) acquired by UAV were processed using Agisoft Metashape Professional Edition 1.5.2 [84], which allows to generate the dense cloud and the orthomosaic of each experimental site. During this process, any ground control points (GCPs) were used due to the irregular and dense canopy cover. The spatial variability in the chestnut orchard was evaluated in terms of vigor and assuming the correspondence between NDVI and vigor [85,86]. NDVI was used as a further filter threshold, as described in Section 2.4.

The dense cloud obtained was normalized using a digital elevation model (DEM) from the automatic classification of ground points from photogrammetric software and subsequently imported into QGis software [87] to develop, by means of the LAStools toolbox [88], the CHM relative to the canopy height of each sample tree. The resolution chosen for this model was 0.05 m.

The next processing step concerned the creation of a chestnut crown mask through a two-fold approach: supervised and unsupervised segmentation. The supervised method consisted of manually drawing each chestnut crown one by one within the experimental plot, visualizing together the CHM and the NRG orthomosaic in the QGis software. The unsupervised approach used a script called 'rLIDAR' (version 0.1.1) [89] in R programming language (version 3.6.0), which allows to generate a vector format file relative to the position and the crown dimension of each sample tree. First, CHM smoothing was performed to eliminate spurious local maxima caused by tree branches. Then, the location and height of individual trees were automatically detected using the CHM and the Local maxima method (rLiDAR: FindTreesCHM function) by sequentially searching the moving window through a Fixed Window Size (FWS) set to 9x9 pixels. In this step, we used a lower CHM resolution (0.25 m/pixel) to generate the mask with the unsupervised method, due to the fact that the workflow with native resolution (0.05 m/pixel) required to many computing resources. However, the segmentation provided enough accuracy with respect to the supervised segmentation. The threshold for the lowest tree height (minht) was fixed at 3.0 m to avoid the misdetection of forest undergrowth as trees. For unsupervised crown segmentation, the ForestCAS function (cf. rLiDAR) based on the watershed method was applied to automatically detect crown boundaries. The threshold for the maximum crown radius (maxcrown) was set to 15.0 m, according to chestnut dendrometric characteristics.

Finally, the obtained dataset was analyzed to perform a spatial estimation of the potential pruning biomass (Figure 2). The tree crown volume was calculated at the pixel level by integrating the volume of all the individual pixels that were positioned below each tree. This choice was made to deal with the irregular shape of every tree and consequently, to reduce the error usually produced in empirical estimations due to the inexact assimilation of trees to regular solids. Therefore, as suggested by Torres-Sanchez et al. [90], the height and area of every tree pixel were multiplied to obtain the pixel volume; subsequently, the crown projected volume was derived by adding the volume of all the pixels below each chestnut tree.

For identifying the volume change between the two years (before and after pruning), the tree mask generated both in the supervised and unsupervised method for the 2017 dataset was chosen as the reference and also used in 2018 (QGis Software) to correct the XY shift of the camera GPS between the two flights. This operation avoided an overestimation of the segmented crown area in 2018 following pruning, especially with R software. The Wpw estimation was performed on the basis of the crown volume reduction in the post-pruning survey with respect to the first flight. A linear regression model between the Wpw measured on the ground and estimated by UAV was applied to evaluate the performance of the UAV approach.

Since the aim of this work was Wpw estimation from remote sensing data, for ground truth measurements, we focused on a very large number of pruning wood sampling (30 trees/site), while only the DBH parameter was meausured as geometric field data for sample trees selection. As a consequence, the geometric data evaluation was related only to the comparison between the supervised and unsupervised methods on the structure from the motion dataset, without any field data as ground truth (tree height or crown dimension).

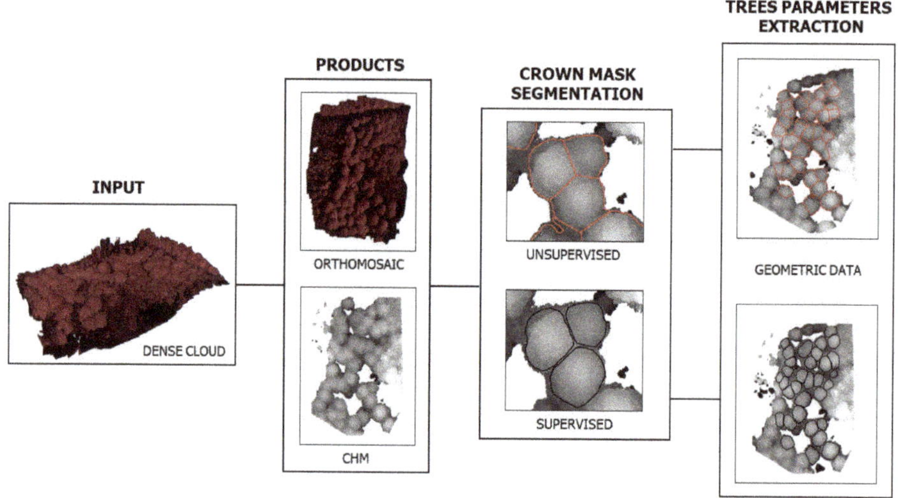

Figure 2. UAV elaboration data workflow.

2.4. Double Filtering Approach

Several authors reported on the improvement in tree crown segmentation when vegetation indices analysis is applied in discriminating between vegetation targets [76,90]. However, in our study, the discrimination between canopy and no canopy pixels was ensured by the CHM thanks to the higher tree height which was more than double the regular plantation sites observed in other works [76,90]. Although the spectral data were not used to improve the segmentation of the crowns from the soil, in our study, they were used to improve the measurement of volume reduction from the CHM.

In detail, the elaboration of crown volume data from the pruned tree (2018 survey) accounted also for no canopy information of the small holes within the canopy undetected by the 3D reconstruction process performed with Agisoft Metashape but clearly visible in the orthomosaic. To solve these problems, we applied a double filtering process: the first one based on canopy height (CHM) and the second based on a vegetation index (NDVI) threshold to remove no vegetation pixels within the crown (Figure 3). The results presented in this work were obtained from a dataset filtered with an NDVI threshold of 0.3.

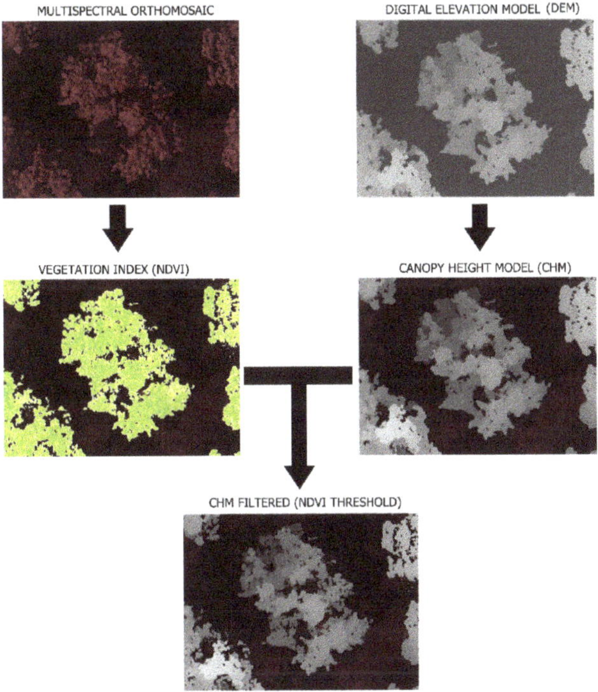

Figure 3. Filtering workflow aimed to improve the accuracy of the canopy height model.

2.5. Puning Wood Biomass Estimation

Reference segmentation masks of sample trees were manually created for each experimental site. The supervised method was applied to develop a linear model between ground truth Wpw and crown projected volume reduction extracted by the reference segmentation mask. The linear model was then applied to calculate the estimated Wpw from the unsupervised segmentation method following Equation (1):

$$Y = \beta\,(X_1 - X_2) + \gamma \tag{1}$$

where Y is the dependent variable (Wpw), X_1 to X_2 are independent variables related, respectively, to crown-projected volume before and after pruning management, β is the multiplicative parameter and γ is the intercept. The coefficient of determination (R^2) and Root Mean Square Error (RMSE) were computed between the measured correlations, between supervised and unsupervised segmentation approaches for UAV geometric data extraction, and between ground-truth measured and estimated Wpw data.

The adjusted coefficient of determination (Equation (2)), the relative root mean square error (Equation (3)) and the percentage bias (Equation (4)) to determine the accuracy of unsupervised segmentation for estimating Wpw using crown projected volume reduction are as follows:

$$adjR2 = 1 - ((n-1)\sum (yini = 1 - \hat{y}i)2(n-p)\sum (yini = 1 - \bar{y}i)2) \tag{2}$$

$$rRMSE = RMSE\bar{y} \tag{3}$$

$$PBias = 100*(\sum (\hat{y}i - yini = 1)n) \tag{4}$$

where n is the number of trees, yi is the field measured Wpw i, $\bar{y}$ is the the mean observed value of Wpw and $\hat{y}i$ is the estimated value of Wpw derived from the linear regression model. A statistical analysis was performed using R software.

3. Results

3.1. Wpw Ground Measurement

In Table 2 are presented the ground data of measured Wpw per tree, per site (sample of 30 trees) and per surface unit.

Table 2. Pruning wood biomass (Wpw) produced per tree (Avg. and Std. dev.), per plot, and per surface.

SiteTitle	DBH (cm)	Wpw per Tree (kg$_{dw}$)	Wpw per Site (kg$_{dw}$)	Wpw per Surface (Mg$_{dw}$ ha^{-1})
A	84.21 ± 26.55	625.01 ± 590.48	18750.28	24.92
B	63.24 ± 11.97	243.69 ± 90.42	7310.85	8.94
C	49.77 ± 11.75	97.37 ± 63.99	3115.84	3.86
D	53.38 ± 15.5	113 ± 76.39	3390.00	4.05

Note: dw = dry weight.

The yield of Wpw per hectare was comparable with the data presented in a former study [91], where three chestnut groves produced from 22 up to 33 Mg ha^{-1}, but this result matches only with site A. In fact, although site A was different from the other three sites presented in the aforementioned study, showing an even bigger DBH on average compared to them, it had a similar pruning intensity. Compared to site A, the number of pruning residues produced in the other three sites investigated in the present study turned out to be noticeably lower, probably caused here—as in other case studies—by differences in trees age, site density, and pruning intensity. The results of sites C and D were very similar and still comparable to site B in terms of wood biomass recovered after pruning, despite the different chestnut varieties.

The proportion of "wood" compared to "wings" was equal to 311.9% at site A, 55.7% at site B and 51.7% at site D. At site C, it was not possible to separate the two fractions due to operative reasons.

3.2. Supervised Data Extraction

Table 3 shows the geometric characterization of each experimental site arising from supervised segmentation before (2017) and after pruning (2018). Maximum tree height, crown mean height, crown area, and crown projected volume are mean values of each site. Crown area and crown projected volume per site are also showed to provide a general overview of biomass reduction after pruning. These values derive from a single tree crown area and projected crown volume, respectively, multiplied by the total number of trees for each site.

The decrease in heights after pruning is not significantly different. However, considering crown mean height and crown area, biomass reduction between the two years detected by UAV is relevant. In fact, it ranges from 8.8% in site A to 14.2% in site B, referring to the crown mean height and from 6.5% in site B to 15.1% in site A, considering the crown area values. In sites C and D, these two parameters show intermediate but comparable variations, reflecting the geomorphological and vegetational similarities of the two sites, in detail: crown mean heights of 11.0% (C) and 11.6% (D), and crown areas of 13.1% (C) and 11.8% (D).

The tree geometric characteristic that best shows the effects of pruning is crown projected volume, whose values have the strongest variations between 2017 and 2018. The highest percentage of biomass reduction was found at site C (21.4%) while site A has the maximum decrease (298 m^3), confirming ground measurements (see Table 2).

Table 3. Geometric characterization of each experimental site.

Year	Site	Tree Height (m)	Crown Mean Height (m)	Crown Area (m^2)	Crown Projected Volume (m^3)	Crown Area per Site (m^2)	Crown Projected Volume per Site (m^3)
2017	A	18.07 ± 2.8	14.69 ± 2.67	93.97 ± 49.56	1401.66 ± 807.53	3006.89	44,853.18
	B	15.19 ± 1.38	12.90 ± 1.09	59.44 ± 20.32	767.61 ± 267.9	1902.10	24,563.55
	C	9.63 ± 0.98	7.65 ± 0.94	61.19 ± 18.69	472.7 ± 170.34	1958.17	15,126.38
	D	11.23 ± 1.36	8.91 ± 1.15	63.06 ± 18.70	575.36 ± 216.96	1954.86	17,836.20
2018	A	17.63 ± 2.66	13.40 ± 2.58	79.74 ± 45.82	1103.71 ± 751.49	2551.80	35,318.81
	B	14.63 ± 1.36	11.07 ± 1.56	55.57 ± 20.56	620.82 ± 248.39	1778.20	19,866.23
	C	9.44 ± 1.12	6.81 ± 1.04	53.16 ± 19.55	371.31 ± 164.53	1700.97	11,882.00
	D	11.08 ± 1.68	7.88 ± 1.33	55.62 ± 22.97	460.28 ± 226.76	1724.20	14,268.58

3.3. Unsupervised Data Extraction

Table 4 reports the segmentation results of the proposed unsupervised methodology. Following Marques et al. [76]'s study, the evaluation of the automatic segmentation accuracy applied in this work was assessed by comparing it with a manual crowns' segmentation. In line with the different site conditions in terms of trees age and dimension, the proposed methodology provides a different response in terms of accuracy. Site A presented the lowest accuracy value (46.7%) due to the highest presence of both over and under detection cases. In detail, the irregular and oversize crown (mean values over 90 m^2) caused elevated crown shape fragmentation (33.3%), while the high overlap crown level led to 20.0% of merged cases. Sites B and D, characterized by a lower overlap level, showed the best accuracy performances, respectively 83.3% and 76.7%. Site D presented a lower accuracy value due to the 20.0% of merged crowns in a circumscribed zone with close trees with similar crown heights. An intermediate accuracy performance was found in the C site (63.3%), where the lowest values and variability in terms of height and some irregular shape cases caused an overestimation of 26.7% crown shape segmentation. The methodology provided the optimal results in terms of undetected tree crown, with 1.7% mean accuracy considering the overall dataset (four sites).

Table 4. Report of the trees' detection accuracy with the number of estimated trees and its detection type in the four sites.

Site	Reference Crowns	Matched	Split	Merged	Missed
A	30	46.7%	33.3%	20.0%	0.0%
B	30	83.3%	3.3%	10.0%	3.3%
C	30	63.3%	26.7%	6.7%	3.3%
D	30	76.7%	3.3%	20.0%	0.0%
Dataset	30	67.5%	16.7%	14.2%	1.7%

3.4. Geometric Data Comparison between the Supervised and the Unsupervised Approach

Figure 4 presents the comparison results between supervised and unsupervised segmentation approaches to perform tree geometric characterization from the structure of motion products. Taking into account the presence of split and merged cases in the unsupervised approach, the dataset was first analyzed not tree-by-tree but by means of the aggregation per site of each polygon identified by both segmentation methodologies. Each XY graph shows the comparison of geometric data related to four sites in both years (2017–2018). The height estimation both for tree height and crown mean height was correctly described from the proposed unsupervised method, providing R^2 = 1.00 correlation coefficient and a good accuracy in terms of values RMSE = 0.25 m and RMSE = 0.24 m, respectively. Considering the estimation of the crown area mean value per site, no correlation was found from the application of the proposed methods (R^2 = 0.01), with a high difference between the values (RMSE = 21.47 m^2). The crown-projected volume shows a lower correlation (R^2 = 0.54) than pure height-derived variables (tree height and crown mean height), but a discrete error in the absolute values (RMSE = 274.64 m^3).

Regarding full site characterization in terms of canopy cover area and crown projected volume, the unsupervised method provided very high correlations: $R^2 = 0.93$ and $R^2 = 0.99$, respectively.

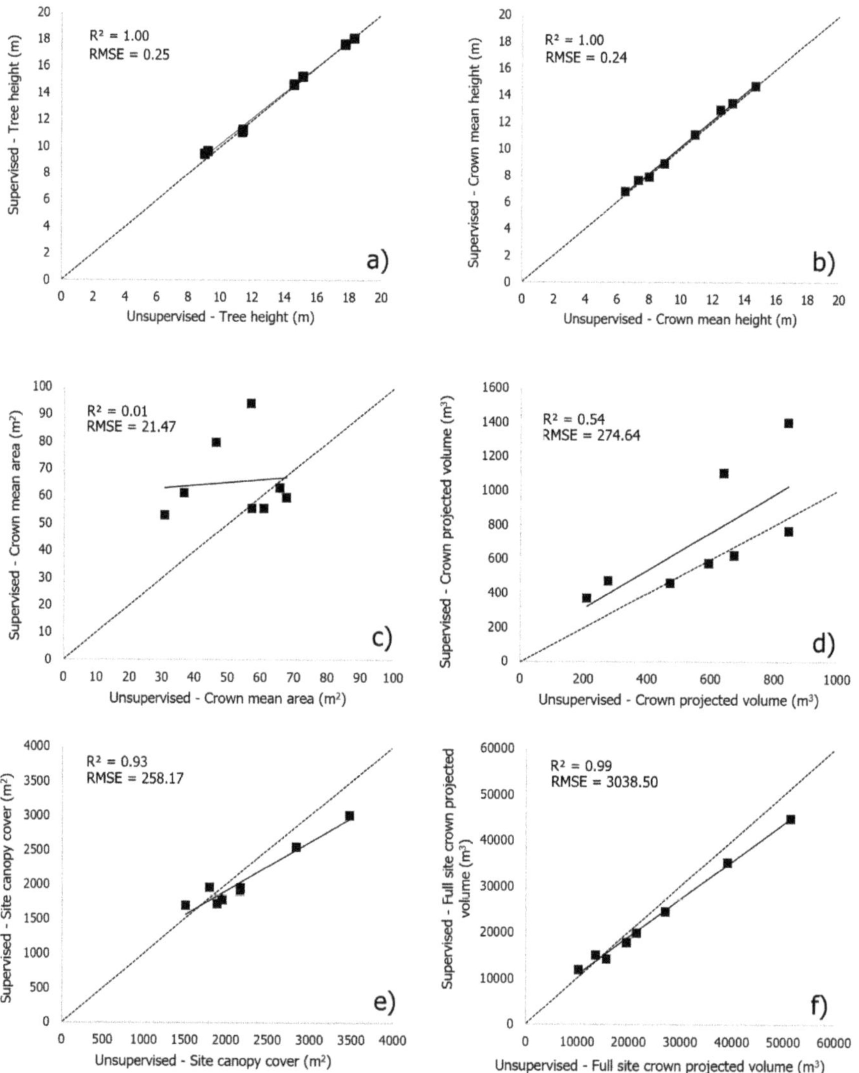

Figure 4. Comparison performance between supervised and unsupervised segmentation within 4 sites for 2017 and 2018 season to characterize geometric information: tree height (**a**), crown mean height (**b**), crown mean area (**c**), crown projected volume (**d**), full site canopy cover (**e**), full site crown projected volume (**f**).

A deep analysis was performed taking into account a larger dataset obtained by the unsupervised segmentation to evaluate the performance of the proposed method as a feasible tool for Wpw evaluation on large scale areas. The unsupervised dataset was created with about 67.5% matched polygons, potentially available to investigate correlation tree-to-tree with measured ground truth Wpw and supervised geometric data per tree. The dataset was then increased by adding the 16.7% of split cases considered as single-tree data by the merging of the sub-polygons in which a sample tree was fragmented by the unsupervised approach. Figure 5 reports the correlations (R^2 and RSME) related to crown volume reduction of the proposed methodology versus the manually segmented mask.

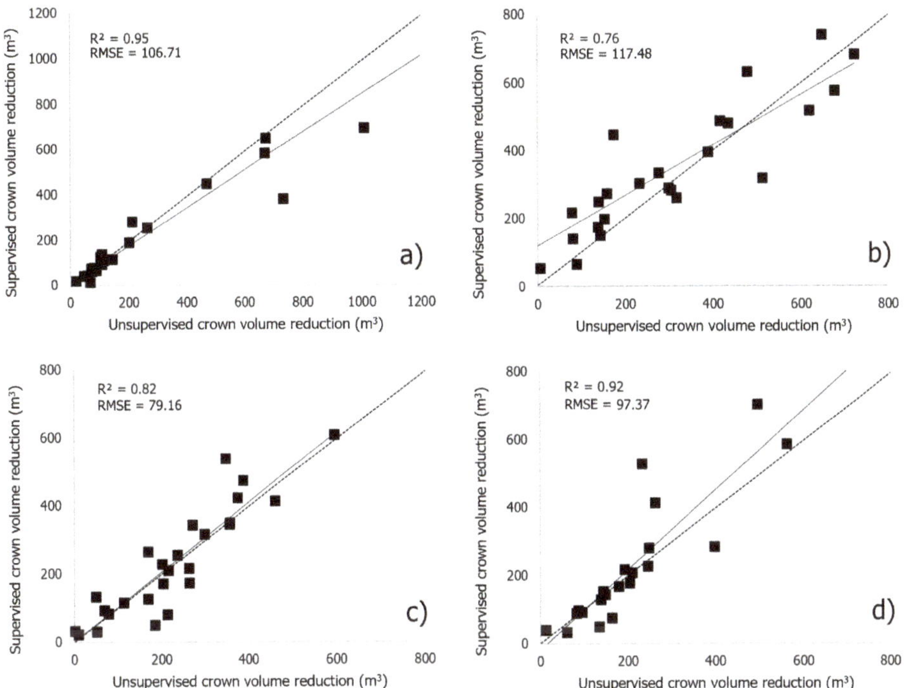

Figure 5. Comparison between crown projected volume reduction extracted by means of the supervised and unsupervised approach within each site (**a**, **b**, **c**, **d**). The dataset was made of both correctly segmented crown (matched) and fragmented crown (split) as a sum of each sub-polygon of the fractionated crown described in Table 4.

The unsupervised method showed a high accuracy performance in crown segmentation, providing high R^2 values ranging between 0.76 and 0.95, and good precision in term of absolute values, with RMSE ranging between 79.16 m^3 and 117.48 m^3. The scatterplots show results close to the 1:1 line between supervised and unsupervised segmentation methods.

3.5. Wpw Estimation

Table 5 presents the correlation results (equations and R^2) between crown projected volume reduction (X-independent variable) and pruning wood biomass (Y-dependent variable), in which a linear regression model was applied to the dataset extracted with the manually reference masks and the ground truth Wpw measurements. All sites show representative results with higher correlation coefficients for the A site ($R^2 = 0.78$), intermediate value in site C and D ($R^2 = 0.71$ and $R^2 = 0.69$ respectively) and lower in site B ($R^2 = 0.60$). Considering the similar tree ages and dimensions in the close sites C and D, Table 5 also reports the good correlations obtained by oganising the two sites as a single dataset ($R^2 = 0.65$). The linear regression analysis applied to the overall dataset provides good results but lower than the values at the single-site level ($R^2 = 0.33$).

Table 5. Regression analysis between ground truth Wpw data (X-independent variable) and estimated projected crown volume reduction (Y-dependent variable) obtained by the application of supervised segmentation. Linear regression results (equation and R^2) calculated for each site (A, B, C, D) and aggregated dataset (C + D and A + B + C + D). All liner regressions provided significance results ($p < 0.001$).

Segmentation	Site	Equation	R^2
Supervised	A	y = 1.2566x − 201.4442	0.78
	B	y = 0.2729x + 143.1937	0.60
	C	y = 0.3549x + 25.1030	0.71
	D	y = 0.2028x + 64.0793	0.69
	C + D	y = 0.2393x + 53.1303	0.65
	A + B + C + D	y = 0.6664x + 56.446	0.33

Concerning the Wpw validation, the estimated Wpw values obtained using the calibration realized in each site with the manually segmented mask were compared with ground-truth Wpw measurements. Figure 6 shows the linear regression results within every single site and with the aggregated dataset (C + D and all dataset), which are similar to the trend found in Table 5.

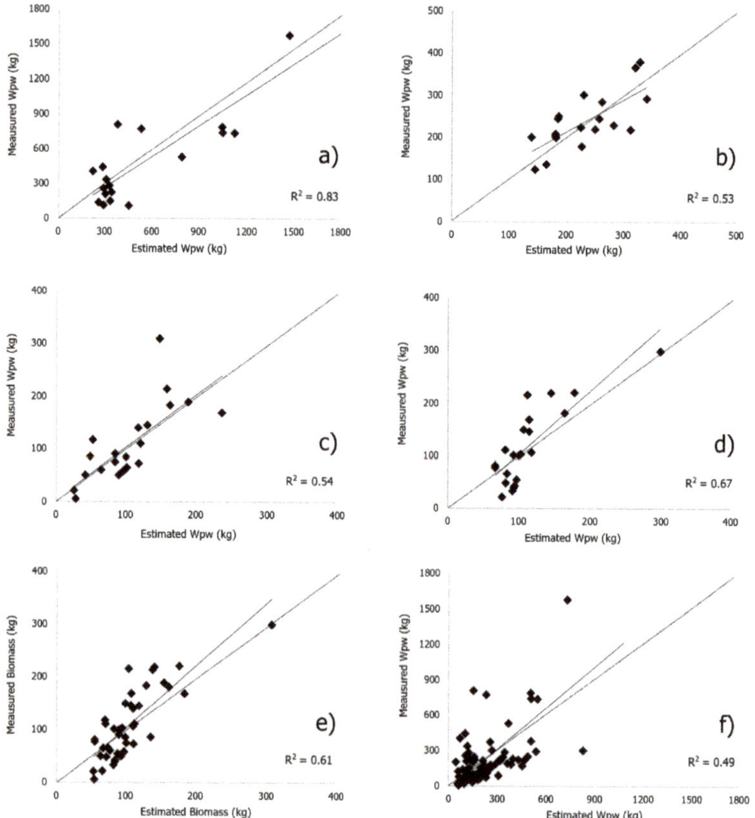

Figure 6. Wpw estimation validation by means of a comparison between measured and estimated Wpw retrieved by the unsupervised segmentation within site A (**a**), B (**b**), C (**c**), D (**d**), C + D (**e**) and A + B + C + D dataset (**f**).

Table 6 reports the statistic results of the methodology suggested as Wpw estimation approach. Site A presents a higher accuracy in Wpw estimation ($R^2 = 0.83$) but RMSE = 221.26 kg is very high, while site B presents lower correlations ($R^2 = 0.53$ and RMSE = 46.99). Sites C and D show good correlations with $R^2 = 0.54$ and $R^2 = 0.67$ respectively, and similar values in terms of RMSE 47.38 kg and 43.08 kg, respectively. Combining the C and D datasets, the results obtained show similar behavior to the separate dataset analysis (with $R^2 = 0.61$ and RMSE = 45.28 kg). The correlations identified using the overall dataset present a lower correlation coefficient ($R^2 = 0.49$) than the other sites, but similar to sites B and C. As reported in Table 6, the unsupervised methodology provides rRMSE values between 44.10% (site A) and 71.60% (site B), and presents very low PBias values with a minimal tendency of understimation in sites B, C, D (mean value −3.0%). Site A presents an overestimation tendency with a bias value of 12.60%.

Table 6. Statistic results (R^2, adjR2, RMSE, rRMSE (%) and PBias (%) calculated for each site (A, B, C, D) and aggregated dataset (C + D and A + B + C + D).

Site	R^2	adjR^2	RMSE	rRMSE (%)	PBias (%)
A	0.83	0.82	221.26	44.10	12.60
B	0.53	0.50	46.99	71.10	−2.50
C	0.54	0.52	47.38	66.20	−2.30
D	0.67	0.65	43.08	58.10	−4.40
C + D	0.61	0.60	45.28	62.70	−3.40
A + B + C + D	0.49	0.48	217.54	71.60	−1.20

4. Discussion

This work aimed to evaluate the accuracy performance of supervised and unsupervised methodologies to estimate pruned biomass. To carry out this objective, an experimental design was planned by taking into account four sites with different conditions of vegetative growth in terms of DBH dimension, trees density and height. Site A presented the highest vegetative growth conditions, site B an intermediate level, C and D, the lowest dimensions. Those different vegetative conditions directly affected the geometric estimation provided by the UAV data analysis, so it was necessary to divide the dataset into three different groups of trees according to age and therefore, size.

Pruning intensity varied among sites as the tender's choice normally depends on the trees' growth, health conditions and age. The heavier intervention carried out with restoration purposes, as in site A, corresponded to a higher wood production, in line with analogous cases. In orchards where pruning is conducted on a long timespan (8–10 years) the amount of wood residues available for industrial purposes can be relevant (from 51.7% to 311.9% in the present study) and an early survey can provide useful information for planning the supply.

The results show that UAV monitoring has a good performance in detecting the biomass reduction after pruning, despite the differences between the trees' geometric characteristics mentioned in Section 3.1. The tree height decrease was weakly detected, mainly due to some branches not being pruned in 2017, which therefore attenuated the height reduction with their vigorous growth in 2018. Considering crown mean height and crown area, the biomass decrease is evident because they are more representative features of the whole canopy. By analyzing the data shown in Table 3, it can be stated that the crown projected volume is the best tree geometric characteristic with which biomass variation can be monitored. In site C, the highest volume decrease is not associated with the greatest height decrease and this can be explained by the typical chestnut pruning method. This technique is not characterized by a uniform topping and hedging but by the cutting of whole branches, so there is no marked height reduction. This led to the formation of crown holes whose presence can be clearly detected only by analyzing tree volume.

In the present work, the chestnut orchard condition strongly affected the segmentation accuracy. The high variability due to irregular spacing between trees, ages and dimensions, irregular crown shape, absence of isolated tree cases, high overlap crown conditions and minimal presence of free space around each tree crown caused a lower accuracy performance in tree segmentation with respect to elevated tree detection results described in other valuable works such as those suggested by Marques et al. [76] and Jorge Torres-Sánchez et al. [90]. The large number of oversized tree crowns in the site A led to an increase in the percentage of split and merged cases, while in sites C and D, the lowest values of DBH and trees' height affect the segmentation accuracy providing 26.7% of split and 20% of merged cases respectively. Concerning the feasibility of large areas canopy cover scouting, the method proposed provided similar results as those reported by Marques et al. [76] related to an undetected tree percentage value (1.7%).

The evaluation of the accuracy of the unsupervised method applied in this study was realized by the comparison of a geometric dataset for each site in each year with the data extracted with the reference mask manually drawn. The proposed method provides a correct estimation of the mean height per tree in terms of tree height with minimal difference in absolute values (RMSE = 0.25 m), confirming the good performance as a height estimation tool for that type of survey on large areas. The mean crown area per tree within each site presents poor correlations and a high RMSE value, as a consequence of the wide value range derived from the split and merge cases of the unsupervised segmentation polygons. The crown projected volume shows a better performance in terms of correlations than the crown area due to the positive influence of the well estimated height but still with a high RMSE value. The analysis of the total canopy cover and the total crown projected volume per site provides optimal results, confirming the method as a powerful tool for fast detection in large areas. A focus elaboration on the projected crown volume reduction between the two years as a consequence of the pruning management practice was performed by increasing the dataset of the matched crowns with the sum of the sub-polygons in which some crowns were divided, reported as "split cases". The improved dataset shows the highest correlation coefficients (mean R^2 = 0.86) and a low difference in values (mean RMSE = 99.75 m^3) with respect to manually segmented crowns.

The validation of the method was carried out after a calibration step, a model was created using a regression analysis between Wpw and volume variation extracted with the reference mask manually drawn on the crown profile. Subsequently, the model identified was applied to the segmentation results obtained with the proposed method. The estimated Wpw per crown was finally correlated with the measured Wpw in order to define the accuracy in terms of correlation coefficient and RMSE. In the case of a full dataset analysis, the approach obtained good correlation (R^2 = 0.33 for calibration and R^2 = 0.49 for validation) but the clustered nature of the dataset with different tree conditions implied a lower performance, confirming the application of a site-by-site approach as the most correct choice. The method proposed showed the high coefficient of correlation (R^2 = 0.83) for site A, but with a very high RMSE since this site was extremely modified in terms of Wpw removed more than the others. Despite showing a lower accuracy in Wpw estimation (R^2 = 0.53), site B presented an acceptable RMSE = 46.99 kg. The factors that strongly reduced the segmentation success were the higher tree density within the site and an elevated overlap level between adjacent crowns, as reported in Table 3. In fact, this site is characterized by high tree and DBH values (close to the biggest in site A), but the lowest mean crown area with respect to all other sites. The application of the proposed methods in sites C and D characterized by similar tree conditions provided high and concordant performances, suggesting that they should be considered as a single dataset (R^2 = 0.61). This result strongly encourages the hypothesis of the feasibility of this method as a site-specific tool for large-scale monitoring. The PBias indicates an overall tendency of minimal understimation, while in site A, characterized by very different conditions in term of Wpw harvested and tree ages, the unsupervised segmentation approach shows a low overstimation of the Wpw with respect to ground truth meausurements.

In the literature, there are no studies regarding chestnut Wpw estimation using UAV. Nevertheless, there have already been recent studies that specifically derived individual biomass and V_{SfM} at tree level with ITC segmentation. Among these, important references in the Mediterranean environment were represented mainly by Guerra-Hernandez et al. [57] and Guerra-Hernandez et al. [56], who used a fixed-wing UAV equipped with an RGB camera to evaluate (i) Wa$_{SfM}$ in *Pinus pinea* regular forest plantation (10 x 16 m regular spaced, open canopy, fairly flat terrain, no understory) and (ii) V$_{SfM}$ in *Eucalyptus* regular forest plantation (3.7 × 2.5 m regularly spaced, steep terrain). Comparing our results with the aforementioned studies, it worth noting that the RMSE could only be compared with Guerra-Hernandez et al. [57], who used the same unit (kg) and reported a value of 87.46 and 117.80 kg for 2015 and 2017, respectively. These values are lower than the overall aggregated dataset RMSE (217.54 kg, Table 6) and the difference could be partially explained by the regular characteristics of the stand investigated by the reference study (spacing, tree age, field management). Thereafter, it is of pivotal importance to compare different remote sensed tree biomasses through statistic indexes that facilitate comparison between datasets or models with different scales, as rRMSE and adjR^2. Guerra-Hernandez et al. [57] in *P. pinea* plantation gained good results in the estimation of Wa$_{SfM}$ in comparison to measured Wa (0.85 < adjR^2 < 0.87 and 11.44% < rRMSE < 12.59% in two different years and model approaches) while Guerra-Hernandez et al. [56], in a *Eucalyptus* plantation, got slightly worse performances (R^2 = 0.43 and rRMSE = 20.31%). However, the current work presents lower correlation values (except in one case) and lower rRMSE (Table 6) with respect to the literature references, mainly due to orchard characteristics (uneven-aged and irregularly spaced) and fine pruning evaluation purposes with respect to growth monitoring. As for Guerra-Hernandez et al. [57], who focused on canopy management study in fruit production crop, our method falls within precision agriculture applications while most of the literature focused on precision forestry.

A strong point of this method was that the dataset was acquired with a low-resolution multispectral camera, which provides both geometric information from the CHM reconstruction and spectral data to calculate the NDVI layer used as a filtering approach to improve the quality of the dataset. As a consequence, the weight of the products to be processed were much lower, allowing faster data processing and requiring less computing power.

5. Conclusion

In the context of the Circular Economy envisaged as a "regenerative system in which resource input and waste, emission, and energy leakage are minimized by slowing, closing and narrowing material and energy loops" [92], it is important to estimate the amounts of wastes available as by-products for industrial purposes. In this specific case, chestnut pruning and their periodical availability can be forecasted and included in supply chain planning to benefit both producers and industrial users.

The unsupervised segmentation method proposed in this work made it possible to realize an accurate estimation of chestnut geometric characteristics from high-resolution CHM layers in four study sites. The results obtained are strongly in line with those extracted with a reference manually segmented mask. Applying a calibration performed on supervised UAV data extraction, the method reports a high accuracy in terms of R^2 and RMSE values, suggesting this approach as a fast and cost-effective tool for fast monitoring of large areas. The dataset was acquired before and after a pruning management practice in four study sites identifying three different DBH classes (around ~0.50 m, ~0.60 m, ~0.80 m). The results obtained allow for us to conclude that the method provides generally good performance, but to achieve the best Wpw estimation, is necessary to choose the correct calibration curve in the function of the DBH. This input information could be easily provided by the orchard owner, making the proposed method a useful tool for fast Wpw estimation purposes.

A future perspective could be to evaluate the potential of a combined approach analyzing also the spectral data actually used only to improve data extraction accuracy, with the aim of finding the described Wpw estimation performance by the contribution of information on vegetation indices.

Author Contributions: C.N. and S.F.D.G. designed the experiment. C.N., S.F.D.G., R.D., L.P., P.T., and A.M. formulated the research methodology and wrote the manuscript. A.B. performed the UAV data acquisition. All Authors reviewed and edited the draft. All authors have read and agreed to the published version of the manuscript.

Funding: This work was funded by Reg. UE n. 1305/2013—PSR 2014/2020—Call for "Integrated Projects of Supply chain (Progetti Integrati di Filiera—PIF) 2015". Administrative Order n. 2359 26/05/2015—P.I.F. n. 3/2015 "VACASTO PLUS". Action 16.2—Project: "OPEN RICCIO".

Acknowledgments: Authors' acknowledgments go to Giovanni Alessandri (Agricis Consulting s.r.l.) and Lorenzo Fazzi (Associazione Castagna del Monte Amiata) and to Niccolò Brachetti Montorselli, professional forester for the field measurements. Special thanks go to orchards' owners Mirco Fazzi and Roberto Ulivieri for the support provided during the Project.

Conflicts of Interest: The authors declare no conflict of interest.

References

1. Gallo, R.; Grigolato, S.; Cavalli, R.; Mazzetto, F. GNSS-based operational monitoring devices for forest logging operation chains. *J. Agric. Eng.* **2013**, *44*, 140–144. [CrossRef]

2. Rossi, F.C.; Fritz, A.; Becker, G. Combining satellite and UAV imagery to delineate forest cover and basal area after mixed-severity fires. *Sustainability* **2018**, *10*, 2227. [CrossRef]

3. Peña, M.A.; Altmann, S.H. Use of satellite-derived hyperspectral indices to identify stress symptoms in an Austrocedrus chilensis forest infested by the aphid Cinara cupressi. *Int. J. Pest Manag.* **2009**, *55*, 197–206. [CrossRef]

4. Paiva, P.F.P.R.; Ruivo, M.D.L.P.; Júnior, O.M.D.S.; Maciel, M.D.N.M.; Braga, T.G.M.; De Andrade, M.M.N.; Junior, P.C.D.S.; Da Rocha, E.S.; De Freitas, T.P.M.; Leite, T.V.D.S.; et al. Deforestation in protect areas in the Amazon: A threat to biodiversity. *Biodivers. Conserv.* **2019**, *29*, 19–38. [CrossRef]

5. Dash, J.P.; Pearse, G.D.; Watt, M.S. UAV multispectral imagery can complement satellite data for monitoring forest health. *Remote Sens.* **2018**, *10*, 1216. [CrossRef]

6. Matese, A.; Toscano, P.; Di Gennaro, S.F.; Genesio, L.; Vaccari, F.P.; Primicerio, J.; Belli, C.; Zaldei, A.; Bianconi, R.; Gioli, B. Intercomparison of UAV, aircraft and satellite remote sensing platforms for precision viticulture. *Remote Sens.* **2015**, *7*, 2971–2990. [CrossRef]

7. Yu, L.; Huang, J.; Zong, S.; Huang, H.; Luo, Y. Detecting shoot beetle damage on Yunnan pine using Landsat time-series data. *Forests* **2018**, *9*, 39.

8. Puliti, S.; Ene, L.T.; Gobakken, T.; Næsset, E. Use of partial-coverage UAV data in sampling for large scale forest inventories. *Remote Sens. Environ.* **2017**, *194*, 115–126. [CrossRef]

9. Fankhauser, K.E.; Strigul, N.S.; Gatziolis, D. Augmentation of traditional forest inventory and Airborne laser scanning with unmanned aerial systems and photogrammetry for forest monitoring. *Remote Sens.* **2018**, *10*, 1562. [CrossRef]

10. Clarivate. Web of Science (All Databases). Available online: http://clarivate.com (accessed on 23 December 2019).

11. Safonova, A.; Tabik, S.; Alcaraz-Segura, D.; Rubtsov, A.; Maglinets, Y.; Herrera, F. Detection of Fir Trees (Abies sibirica) Damaged by the Bark Beetle in Unmanned Aerial Vehicle Images with Deep Learning. *Remote Sens.* **2019**, *11*, 643. [CrossRef]

12. Swinfield, T.; Lindsell, J.A.; Williams, J.V.; Harrison, R.D.; Agustiono; Habibi; Gemita, E.; Schönlieb, C.B.; Coomes, D.A. Accurate Measurement of Tropical Forest Canopy Heights and Aboveground Carbon Using Structure from Motion. *Remote Sens.* **2019**, *11*, 928. [CrossRef]

13. Roşca, S.; Suomalainen, J.; Bartholomeus, H.; Herold, M. Comparing terrestrial laser scanning and unmanned aerial vehicle structure from motion to assess top of canopy structure in tropical forests. *Interface Focus* **2018**, *8*, 20170038. [CrossRef]

14. Liang, X.; Wang, Y.; Pyörälä, J.; Lehtomäki, M.; Yu, X.; Kaartinen, H.; Kukko, A.; Honkavaara, E.; Issaoui, A.E.I.; Nevalainen, O.; et al. Forest in situ observations using unmanned aerial vehicle as an alternative of terrestrial measurements. *For. Ecosyst.* **2019**, *6*, 20. [CrossRef]

15. Maes, W.H.; Huete, A.R.; Avino, M.; Boer, M.M.; Dehaan, R.; Pendall, E.; Griebel, A.; Steppe, K. Can UAV-based infrared thermography be used to study plant-parasite interactions between mistletoe and Eucalypt trees? *Remote Sens.* **2018**, *10*, 2062. [CrossRef]

16. Grznárová, A.; Mokroš, M.; Surový, P.; Slavík, M.; Pondelík, M.; Mergani, J. The crown diameter estimation from fixed wing type of uav imagery. *Int. Arch. Photogramm. Remote Sens. Spat. Inf. Sci. ISPRS Arch.* **2019**, *42*, 337–341. [CrossRef]

17. Pádua, L.; Adão, T.; Guimarães, N.; Sousa, A.; Peres, E.; Sousa, J.J. Post-fire forestry recovery monitoring using high-resolution multispectral imagery from unmanned aerial vehicles. *Int. Arch. Photogramm. Remote Sens. Spat. Inf. Sci. ISPRS Arch.* **2019**, *42*, 301–305. [CrossRef]

18. Puliti, S.; Saarela, S.; Gobakken, T.; Ståhl, G.; Næsset, E. Combining UAV and Sentinel-2 auxiliary data for forest growing stock volume estimation through hierarchical model-based inference. *Remote Sens. Environ.* **2018**, *204*, 485–497. [CrossRef]

19. Aguilar, F.J.; Rivas, J.R.; Nemmaoui, A.; Peñalver, A.; Aguilar, M.A. UAV-based digital terrain model generation under leaf-off conditions to support teak plantations inventories in tropical dry forests. A case of the coastal region of Ecuador. *Sensors* **2019**, *19*, 1934. [CrossRef]

20. Iizuka, K.; Kato, T.; Silsigia, S.; Soufiningrum, A.Y.; Kozan, O. Estimating and examining the sensitivity of different vegetation indices to fractions of vegetation cover at different scaling Grids for Early Stage Acacia Plantation Forests Using a Fixed-Wing UAS. *Remote Sens.* **2019**, *11*, 1816. [CrossRef]

21. Zou, X.; Liang, A.; Wu, B.; Su, J.; Zheng, R.; Li, J. UAV-based high-throughput approach for fast growing Cunninghamia lanceolata (Lamb.) cultivar screening by machine learning. *Forests* **2019**, *10*, 815. [CrossRef]

22. Frey, J.; Kovach, K.; Stemmler, S.; Koch, B. UAV photogrammetry of forests as a vulnerable process. A sensitivity analysis for a structure from motion RGB-image pipeline. *Remote Sens.* **2018**, *10*, 912. [CrossRef]

23. Hentz, Â.M.K.; Silva, C.A.; Dalla Corte, A.P.; Netto, S.P.; Strager, M.P.; Klauberg, C. Estimating forest uniformity in Eucalyptus spp. and Pinus taeda L. stands using field measurements and structure from motion point clouds generated from unmanned aerial vehicle (UAV) data collection. *For. Syst.* **2018**, *27*, 1–17. [CrossRef]

24. Liu, K.; Shen, X.; Cao, L.; Wang, G.; Cao, F. Estimating forest structural attributes using UAV-LiDAR data in Ginkgo plantations. *ISPRS J. Photogramm. Remote Sens.* **2018**, *146*, 465–482. [CrossRef]

25. Giannetti, F.; Chirici, G.; Gobakken, T.; Næsset, E.; Travaglini, D.; Puliti, S. A new approach with DTM-independent metrics for forest growing stock prediction using UAV photogrammetric data. *Remote Sens. Environ.* **2018**, *213*, 195–205. [CrossRef]

26. Fraser, B.T.; Congalton, R.G. Evaluating the effectiveness of Unmanned Aerial Systems (UAS) for collecting thematic map accuracy assessment reference data in New England forests. *Forests* **2019**, *10*, 24. [CrossRef]

27. Liebermann, H.; Strager, M.P. Using Unmanned Aerial Systems for Deriving Forest Stand Characteristics in Mixed Hardwoods of West Virginia Using. *J. Geospatial Appl. Nat. Resour.* **2018**, *2*, 1–31.

28. Kattenborn, T.; Lopatin, J.; Förster, M.; Braun, A.C.; Fassnacht, F.E. UAV data as alternative to field sampling to map woody invasive species based on combined Sentinel-1 and Sentinel-2 data. *Remote Sens. Environ.* **2019**, *227*, 61–73. [CrossRef]

29. Tuominen, S.; Balazs, A.; Honkavaara, E.; Pölönen, I.; Saari, H.; Hakala, T.; Viljanen, N. Hyperspectral UAV-Imagery and photogrammetric canopy height model in estimating forest stand variables. *Silva Fenn.* **2017**, *51*, 1–21. [CrossRef]

30. Park, J.Y.; Muller-Landau, H.C.; Lichstein, J.W.; Rifai, S.W.; Dandois, J.P.; Bohlman, S.A. Quantifying leaf phenology of individual trees and species in a tropical forest using unmanned aerial vehicle (UAV) images. *Remote Sens.* **2019**, *11*, 1534. [CrossRef]

31. Chung, C.H.; Wang, C.H.; Hsieh, H.C.; Huang, C.Y. Comparison of forest canopy height profiles in a mountainous region of Taiwan derived from airborne lidar and unmanned aerial vehicle imagery. *GISci. Remote Sens.* **2019**, *56*, 1289–1304. [CrossRef]

32. Carr, J.C.; Slyder, J.B. Individual tree segmentation from a leaf-off photogrammetric point cloud. *Int. J. Remote Sens.* **2018**, *39*, 5195–5210. [CrossRef]

33. Jayathunga, S.; Owari, T.; Tsuyuki, S. Digital Aerial Photogrammetry for Uneven-Aged Forest Management: Assessing the Potential to Reconstruct Canopy Structure and Estimate Living Biomass. *Remote Sens.* **2019**, *11*, 338. [CrossRef]

34. Vaglio Laurin, G.; Ding, J.; Disney, M.; Bartholomeus, H.; Herold, M.; Papale, D.; Valentini, R. Tree height in tropical forest as measured by different ground, proximal, and remote sensing instruments, and impacts on above ground biomass estimates. *Int. J. Appl. Earth Obs. Geoinf.* **2019**, *82*, 101899. [CrossRef]

35. De Luca, G.; Silva, J.M.N.; Cerasoli, S.; Araújo, J.; Campos, J.; Di Fazio, S.; Modica, G. Object-based land cover classification of cork oak woodlands using UAV imagery and Orfeo Toolbox. *Remote Sens.* **2019**, *11*, 1238. [CrossRef]

36. Carvajal-Ramírez, F.; Serrano, J.M.P.R.; Agüera-Vega, F.; Martínez-Carricondo, P. A comparative analysis of phytovolume estimation methods based on UAV-photogrammetry and multispectral imagery in a mediterranean forest. *Remote Sens.* **2019**, *11*, 2579. [CrossRef]

37. Zhang, D.; Liu, J.; Ni, W.; Sun, G.; Zhang, Z.; Liu, Q.; Wang, Q. Estimation of Forest Leaf Area Index Using Height and Canopy Cover Information Extracted From Unmanned Aerial Vehicle Stereo Imagery. *IEEE J. Sel. Top. Appl. Earth Obs. Remote Sens.* **2019**, *12*, 471–481. [CrossRef]

38. McClelland, M.P.; van Aardt, J.; Hale, D. Manned aircraft versus small unmanned aerial system—Forestry remote sensing comparison utilizing lidar and structure-from-motion for forest carbon modeling and disturbance detection. *J. Appl. Remote Sens.* **2019**, *14*, 1. [CrossRef]

39. MacDicken, K.G. Global Forest Resources Assessment 2015: What, why and how? *For. Ecol. Manag.* **2015**, *352*, 3–8. [CrossRef]

40. Waite, C.E.; van der Heijden, G.M.F.; Field, R.; Boyd, D.S. A view from above: Unmanned aerial vehicles (UAVs) provide a new tool for assessing liana infestation in tropical forest canopies. *J. Appl. Ecol.* **2019**, *56*, 902–912. [CrossRef]

41. Wu, Z.; Ni, M.; Hu, Z.; Wang, J.; Li, Q.; Wu, G. Mapping invasive plant with UAV-derived 3D mesh model in mountain area—A case study in Shenzhen Coast, China. *Int. J. Appl. Earth Obs. Geoinf.* **2019**, *77*, 129–139. [CrossRef]

42. Franklin, S.E.; Ahmed, O.S. Deciduous tree species classification using object-based analysis and machine learning with unmanned aerial vehicle multispectral data. *Int. J. Remote Sens.* **2018**, *39*, 5236–5245. [CrossRef]

43. Komárek, J.; Klouček, T.; Prošek, J. The potential of Unmanned Aerial Systems: A tool towards precision classification of hard-to-distinguish vegetation types? *Int. J. Appl. Earth Obs. Geoinf.* **2018**, *71*, 9–19. [CrossRef]

44. Sothe, C.; Dalponte, M.; de Almeida, C.M.; Schimalski, M.B.; Lima, C.L.; Liesenberg, V.; Miyoshi, G.T.; Tommaselli, A.M.G. Tree species classification in a highly diverse subtropical forest integrating UAV-based photogrammetric point cloud and hyperspectral data. *Remote Sens.* **2019**, *11*, 1338. [CrossRef]

45. Dash, J.P.; Watt, M.S.; Paul, T.S.H.; Morgenroth, J.; Pearse, G.D. Early detection of invasive exotic trees using UAV and manned aircraft multispectral and LiDAR Data. *Remote Sens.* **2019**, *11*, 1812. [CrossRef]

46. Wallace, L.; Bellman, C.; Hally, B.; Hernandez, J.; Jones, S.; Hillman, S. Assessing the ability of image based point clouds captured from a UAV to measure the terrain in the presence of canopy cover. *Forests* **2019**, *10*, 284. [CrossRef]

47. Seifert, E.; Seifert, S.; Vogt, H.; Drew, D.; van Aardt, J.; Kunneke, A.; Seifert, T. Influence of drone altitude, image overlap, and optical sensor resolution on multi-view reconstruction of forest images. *Remote Sens.* **2019**, *11*, 1252. [CrossRef]

48. Brach, M.; Chan, J.C.W.; Szymański, P. Accuracy assessment of different photogrammetric software for processing data from low-cost UAV platforms in forest conditions. *IForest* **2019**, *12*, 435–441. [CrossRef]

49. Tomaštík, J.; Mokroš, M.; Surový, P.; Grznárová, A.; Merganič, J. UAV RTK/PPK method-An optimal solution for mapping inaccessible forested areas? *Remote Sens.* **2019**, *11*, 721. [CrossRef]

50. Otsu, K.; Pla, M.; Vayreda, J.; Brotons, L. Calibrating the severity of forest defoliation by pine processionary moth with landsat and UAV imagery. *Sensors* **2018**, *18*, 3278. [CrossRef]

51. Cardil, A.; Otsu, K.; Pla, M.; Silva, C.A.; Brotons, L. Quantifying pine processionary moth defoliation in a pine-oak mixed forest using unmanned aerial systems and multispectral imagery. *PLoS ONE* **2019**, *14*, e0213027. [CrossRef]

52. Zhang, N.; Zhang, X.; Yang, G.; Zhu, C.; Huo, L.; Feng, H. Assessment of defoliation during the Dendrolimus tabulaeformis Tsai et Liu disaster outbreak using UAV-based hyperspectral images. *Remote Sens. Environ.* **2018**, *217*, 323–339. [CrossRef]

53. Larrinaga, A.R.; Brotons, L. Greenness Indices from a Low-Cost UAV Imagery as Tools for Monitoring Post-Fire Forest Recovery. *Drones* **2019**, *3*, 6. [CrossRef]

54. Abdollahnejad, A.; Panagiotidis, D.; Surovỳ, P. Estimation and extrapolation of tree parameters using spectral correlation between UAV and Pléiades data. *Forests* **2018**, *9*, 85. [CrossRef]

55. Domingo, D.; Ørka, H.O.; Næsset, E.; Kachamba, D.; Gobakken, T. Effects of UAV image resolution, camera type, and image overlap on accuracy of biomass predictions in a tropicalwoodland. *Remote Sens.* **2019**, *11*, 948. [CrossRef]

56. Guerra-Hernández, J.; Cosenza, D.N.; Cardil, A.; Silva, C.A.; Botequim, B.; Soares, P.; Silva, M.; González-Ferreiro, E.; Díaz-Varela, R.A. Predicting growing stock volume of eucalyptus plantations using 3-D point clouds derived from UAV imagery and ALS data. *Forests* **2019**, *10*, 905. [CrossRef]

57. Guerra-Hernández, J.; González-Ferreiro, E.; Monleón, V.J.; Faias, S.P.; Tomé, M.; Díaz-Varela, R.A. Use of multi-temporal UAV-derived imagery for estimating individual tree growth in Pinus pinea stands. *Forests* **2017**, *8*, 300. [CrossRef]

58. Kachamba, D.J.; Ørka, H.O.; Næsset, E.; Eid, T.; Gobakken, T. Influence of plot size on efficiency of biomass estimates in inventories of dry tropical forests assisted by photogrammetric data from an unmanned aircraft system. *Remote Sens.* **2017**, *9*, 610. [CrossRef]

59. Li, Z.; Zan, Q.; Yang, Q.; Zhu, D.; Chen, Y.; Yu, S. Remote estimation of mangrove aboveground carbon stock at the species level using a low-cost unmanned aerial vehicle system. *Remote Sens.* **2019**, *11*, 1018. [CrossRef]

60. Sandino, J.; Pegg, G.; Gonzalez, F.; Smith, G. Aerial mapping of forests affected by pathogens using UAVs, hyperspectral sensors, and artificial intelligence. *Sensors* **2018**, *18*, 944. [CrossRef]

61. White, J.C.; Wulder, M.A.; Varhola, A.; Vastaranta, M.; Coops, N.C.; Cook, B.D.; Pitt, D.; Woods, M. A best practices guide for generating forest inventory attributes from airborne laser scanning data using an area-based approach. *For. Chron.* **2013**, *89*, 722–723. [CrossRef]

62. Lindberg, E.; Holmgren, J. Individual Tree Crown Methods for 3D Data from Remote Sensing. *Curr. For. Reports* **2017**, *3*, 19–31. [CrossRef]

63. Alonzo, M.; Andersen, H.E.; Morton, D.C.; Cook, B.D. Quantifying boreal forest structure and composition using UAV structure from motion. *Forests* **2018**, *9*, 119. [CrossRef]

64. Fujimoto, A.; Haga, C.; Matsui, T.; Machimura, T.; Hayashi, K.; Sugita, S.; Takagi, H. An end to end process development for UAV-SfM based forest monitoring: Individual tree detection, species classification and carbon dynamics simulation. *Forests* **2019**, *10*, 680. [CrossRef]

65. Wu, X.; Shen, X.; Cao, L.; Wang, G.; Cao, F. Assessment of individual tree detection and canopy cover estimation using unmanned aerial vehicle based light detection and ranging (UAV-LiDAR) data in planted forests. *Remote Sens.* **2019**, *11*, 908. [CrossRef]

66. Tian, J.; Dai, T.; Li, H.; Liao, C.; Teng, W.; Hu, Q.; Ma, W.; Xu, Y. A novel tree height extraction approach for individual trees by combining TLS and UAV image-based point cloud integration. *Forests* **2019**, *10*, 537. [CrossRef]

67. Saarinen, N.; Vastaranta, M.; Näsi, R.; Rosnell, T.; Hakala, T.; Honkavaara, E.; Wulder, M.A.; Luoma, V.; Tommaselli, A.M.G.; Imai, N.N.; et al. Assessing biodiversity in boreal forests with UAV-based photogrammetric point clouds and hyperspectral imaging. *Remote Sens.* **2018**, *10*, 338. [CrossRef]

68. Brieger, F.; Herzschuh, U.; Pestryakova, L.A.; Bookhagen, B.; Zakharov, E.S.; Kruse, S. Advances in the derivation of Northeast Siberian forest metrics using high-resolution UAV-based photogrammetric point clouds. *Remote Sens.* **2019**, *11*, 1447. [CrossRef]

69. Brovkina, O.; Cienciala, E.; Surový, P.; Janata, P. Unmanned aerial vehicles (UAV) for assessment of qualitative classification of Norway spruce in temperate forest stands. *Geo Spatial Inf. Sci.* **2018**, *21*, 12–20. [CrossRef]

70. Morales, G.; Kemper, G.; Sevillano, G.; Arteaga, D.; Ortega, I.; Telles, J. Automatic segmentation of Mauritia flexuosa in unmanned aerial vehicle (UAV) imagery using deep learning. *Forests* **2018**, *9*, 736. [CrossRef]

71. Yan, W.; Guan, H.; Cao, L.; Yu, Y.; Gao, S.; Lu, J.Y. An automated hierarchical approach for three-dimensional segmentation of single trees using UAV LiDAR data. *Remote Sens.* **2018**, *10*, 1999. [CrossRef]

72. Ganz, S.; Käber, Y.; Adler, P. Measuring tree height with remote sensing-a comparison of photogrammetric and LiDAR data with different field measurements. *Forests* **2019**, *10*, 694. [CrossRef]

73. Balsi, M.; Esposito, S.; Fallavollita, P.; Nardinocchi, C. Single-tree detection in high-density LiDAR data from UAV-based survey. *Eur. J. Remote Sens.* **2018**, *51*, 679–692. [CrossRef]

74. Huang, H.; Li, X.; Chen, C. Individual tree crown detection and delineation from very-high-resolution UAV images based on bias field and marker-controlled watershed segmentation algorithms. *IEEE J. Sel. Top. Appl. Earth Obs. Remote Sens.* **2018**, *11*, 2253–2262. [CrossRef]

75. Mayr, M.J.; Malß, S.; Ofner, E.; Samimi, C. Disturbance feedbacks on the height of woody vegetation in a savannah: A multi-plot assessment using an unmanned aerial vehicle (UAV). *Int. J. Remote Sens.* **2018**, *39*, 4761–4785. [CrossRef]

76. Marques, P.; Pádua, L.; Adão, T.; Hruška, J.; Peres, E.; Sousa, A.; Sousa, J.J. UAV-based automatic detection and monitoring of chestnut trees. *Remote Sens.* **2019**, *11*, 855. [CrossRef]

77. Apostol, B.; Petrila, M.; Lorenţ, A.; Ciceu, A.; Gancz, V.; Badea, O. Species discrimination and individual tree detection for predicting main dendrometric characteristics in mixed temperate forests by use of airborne laser scanning and ultra-high-resolution imagery. *Sci. Total Environ.* **2020**, *698*, 134074. [CrossRef]

78. Tabacchi, G.; De Natale, F.; Di Cosmo, L.; Floris, A.; Gagliano, C.; Gasparini, P.; Genchi, L.; Scrinzi, G.; Tosi, V. *INFC—Le Stime di Superficie 2005, Parte Prima*; CRA—Istituto Sperimentale per l'Assestamento Forestale e per l'Alpicoltura: Trento, Italy, 2007; pp. 1–389.

79. Pádua, L.; Hruška, J.; Bessa, J.; Adão, T.; Martins, L.M.; Gonçalves, J.A.; Peres, E.; Sousa, A.M.R.; Castro, J.P.; Sousa, J.J. Multi-temporal analysis of forestry and coastal environments using UASs. *Remote Sens.* **2018**, *10*, 24. [CrossRef]

80. Pádua, L.; Adão, T.; Hruška, J.; Sousa, J.J.; Peres, E.; Morais, R.; Sousa, A. Very high resolution aerial data to support multi-temporal precision agriculture information management. *Procedia Comput. Sci.* **2017**, *121*, 407–414. [CrossRef]

81. Adão, T.; Pádua, L.; Pinho, T.M.; Hruška, J.; Sousa, A.; Sousa, J.J.; Morais, R.; Peres, E. Multi-purpose chestnut clusters detection using deep learning: A preliminary approach. *Int. Arch. Photogramm. Remote Sens. Spat. Inf. Sci. ISPRS Arch.* **2019**, *42*, 1–7. [CrossRef]

82. Lee, C. A Study on the Use of Unmanned Aerial Vehicles for Chestnut Insect Pests Control. *J. Agric. Life Sci.* **2018**, *52*, 25–29. [CrossRef]

83. Matese, A.; Di Gennaro, S.F. Practical applications of a multisensor UAV platform based on multispectral, thermal and RGB high resolution images in precision viticulture. *Agriculture* **2018**, *8*, 116. [CrossRef]

84. Agisoft, L.C.C. Metashape Professional, Edition 1.5.2, St. Petersburg, Russia. Available online: https://www.agisoft.com/ (accessed on 23 December 2019).

85. Costa Ferreira, A.-M.; Germain, C.; Homayouni, S.; Da Costa, J.-P.; Grenier, G.; Marguerit, E.; Roby, J.-P. Transformation of high resolution aerial images in vine vigour maps at intra-block scale by semi automatic image processing. In Proceedings of the XVth International GESCO Meeting, Porec, Croatia, 20–23 June 2007; pp. 1–7.

86. Fiorillo, E.; Crisci, A.; De Filippis, T.; Di Gennaro, S.F.; Di Blasi, S.; Matese, A.; Primicerio, J.; Vaccari, F.P. Airborne high-resolution images for grape classification: Changes in correlation between technological and late maturity in a Sangiovese vineyard in Central Italy. *Aust. J. Grape Wine Res.* **2012**, *18*, 80–90. [CrossRef]

87. QGIS. Noosa Version. Available online: https://www.qgis.org/it/site/ (accessed on 23 December 2019).

88. Isenburg, M. LAStools—Efficient Tools for LiDAR Processing. Available online: https://www.cs.unc.edu/~{}isenburg/lastools/ (accessed on 23 December 2019).

89. Silva, C.A.; Crookston, N.L.; Hudak, A.T.; Vierling, L.A. Package "rLiDAR". LiDAR Data Processing and Visualization. 2017. Available online: https://cran.r-project.org/web/packages/rLiDAR/index.html (accessed on 23 December 2019).

90. Torres-Sánchez, J.; López-Granados, F.; Serrano, N.; Arquero, O.; Peña, J.M. High-throughput 3-D monitoring of agricultural-tree plantations with Unmanned Aerial Vehicle (UAV) technology. *PLoS ONE* **2015**, *10*, e0130479. [CrossRef] [PubMed]

91. Nati, C.; Montorselli, N.B.; Olmi, R. Wood biomass recovery from chestnut orchards: Results from a case study. *Agrofor. Syst.* **2018**, *92*, 1129–1143. [CrossRef]

92. Geissdoerfer, M.; Savaget, P.; Bocken, N.M.P.; Hultink, E.J. The Circular Economy e A new sustainability paradigm? *J. Clean. Prod.* **2017**, *143*, 757–768. [CrossRef]

 *forests*

Article

Using UAV Multispectral Images for Classification of Forest Burn Severity—A Case Study of the 2019 Gangneung Forest Fire

Jung-il Shin [1,*], Won-woo Seo [2], Taejung Kim [2], Joowon Park [3] and Choong-shik Woo [4,*]

[1] Geoinformatic Engineering Research Center, Inha University, Incheon 22212, Korea
[2] Department of Geoinformatic Engineering, Inha University, Incheon 22212, Korea; winter1260@inha.edu (W.-w.S.); tezid@inha.ac.kr (T.K.)
[3] School of Forest Science & Landscape Architecture, Kyungpook National University, Daegu 41566, Korea; joowon72@knu.ac.kr
[4] Forest Disaster Management Division, National Institute of Forest Science, Seoul 02455, Korea
* Correspondence: jishin@inha.ac.kr (J.-i.S.); woocs@korea.kr (C.-s.W.)

Received: 23 September 2019; Accepted: 11 November 2019; Published: 14 November 2019

Abstract: Unmanned aerial vehicle (UAV)-based remote sensing has limitations in acquiring images before a forest fire, although burn severity can be analyzed by comparing images before and after a fire. Determining the burned surface area is a challenging class in the analysis of burn area severity because it looks unburned in images from aircraft or satellites. This study analyzes the availability of multispectral UAV images that can be used to classify burn severity, including the burned surface class. RedEdge multispectral UAV image was acquired after a forest fire, which was then processed into a mosaic reflectance image. Hundreds of samples were collected for each burn severity class, and they were used as training and validation samples for classification. Maximum likelihood (MLH), spectral angle mapper (SAM), and thresholding of a normalized difference vegetation index (NDVI) were used as classifiers. In the results, all classifiers showed high overall accuracy. The classifiers also showed high accuracy for classification of the burned surface, even though there was some confusion among spectrally similar classes, unburned pine, and unburned deciduous. Therefore, multispectral UAV images can be used to analyze burn severity after a forest fire. Additionally, NDVI thresholding can also be an easy and accurate method, although thresholds should be generalized in the future.

Keywords: UAV; multispectral image; forest fire; burn severity; classification

---

1. Introduction

A fire is a primary disaster in forests, disturbing biodiversity and forest wealth. Forest fires sometimes destroy human settlements and cause loss of life and property. The forest fires in South Korea occur mainly in the dry season (from winter to spring), and are mostly caused by humans. As a forest fire burns off vegetation, soil, organic matter, and moisture, there is a danger of landslides or other secondary disasters during the summer rainy season. In the Republic of Korea, there were 6,588 forest fires from 2004 to 2018. The total area affected was 11,065 hectares, and the damage amounted to US$ 252 million [1].

A strategy is needed to recover from the damage and to respond to secondary disasters by rapidly investigating the burn severity. Burn severity is mainly investigated by field survey or visual interpretation of satellite imagery. Field surveys need a lot of labor, incur high costs, and take time. Satellite imagery has limited uses based on weather conditions and image resolution. Therefore, a rapid and efficient method is needed to investigate burn severity. The unmanned aerial vehicle (UAV) is widely used in various fields. UAVs and sensors provide high-resolution data when users want them, and they are less affected by atmospheric conditions [2–5]. In most cases, UAVs can acquire images right after a forest fire even though the location and time of a forest fire cannot be anticipated [6,7].

Previous studies used spaceborne or airborne multispectral imagery to analyze burn severity. The traditional methods are comparisons of spectral indices pre- and post-fire [8,9]. The normalized difference vegetation index (NDVI) and the normalized burn ratio (NBR) are well-known as spectral indices sensitive to forest fire damage [10–13]. Recent studies widely used NBR and the burned area index (BAI) because shortwave infrared (SWIR) bands are more sensitive to forest fire damage [14–16]. UAV or high-resolution satellite images have to use visible-near infrared (VNIR) bands because they do not have SWIR bands.

Burn severity incorporates both short- and long-term post-fire effects on the local and regional environment. Burn severity is defined as the degree to which an ecosystem has changed as a result of the fire. Vegetation rehabilitation may specifically vary based on burn severity after a fire [17–22]. Previous studies classified burn severity into four or five classes, such as extreme, high, moderate, low, and unburned, using remote sensing data based on the composite burn index (CBI) suggested by the United States Forest Service [23–25]. Those classes might not be clear enough to define burn severity with remote sensing data. One study suggested a Korean CBI (KCBI) by adjusting the CBI classes to burned crown, boiled crown, moderate (a mix of burned crown and burned surface), low (burned surface only), and unburned [26]. But *low* and *unburned* are challenging classes. They look similar from nadir views of the crown because the class *low* means a burned surface under an unburned crown. Also, the Korean forest has a very high canopy density. These characteristics add limitations to classifying severity as low or unburned. Therefore, a method is needed to classify low and unburned severity using remotely sensed imagery that might contribute to estimating damaged areas and establishing a recovery plan.

This study tries to analyze multispectral UAV images that can be used to classify burn severity, including surfaces in the study area that are classed as low under the KCBI. Sample pixels were collected from UAV multispectral images based on visual interpretation and field surveys. Spectral characteristics of the samples were analyzed, after which burn severity was classified using a spectral index and supervised classifiers. The suitability of multispectral UAV imaging for burn severity analysis is shown by the classification accuracy.

2. Study Area and Data

2.1. Study Area

The study area is in a forest near the city of Gangneung, Republic of Korea, which is located on the coast of the East Sea, as seen in Figure 1a. The area is very dry from winter to spring owing to a föhn wind and low precipitation. The Korean red pine (*Pinus densiflora*) is the primary species in the area, which has a volatile pine resin. These environmental and climatic factors are the cause of frequent and huge forest fires. The study area was some of the damaged area from a forest fire that occurred from 4 to 5 April 2019. Seven hundred hectares burned, and US$ 61 million was lost from this forest fire. Figure 1b shows a KOMPSAT-3A high-resolution satellite image taken on 5 April 2019 when the forest fire was in progress, which has a spatial resolution of 2.2 m and four bands (blue, green, red, and near-infrared). The red box is the study area, which is 2 km × 0.5 km in size, located at the border of the whole damaged area, where there were various KCBI types of burn severity.

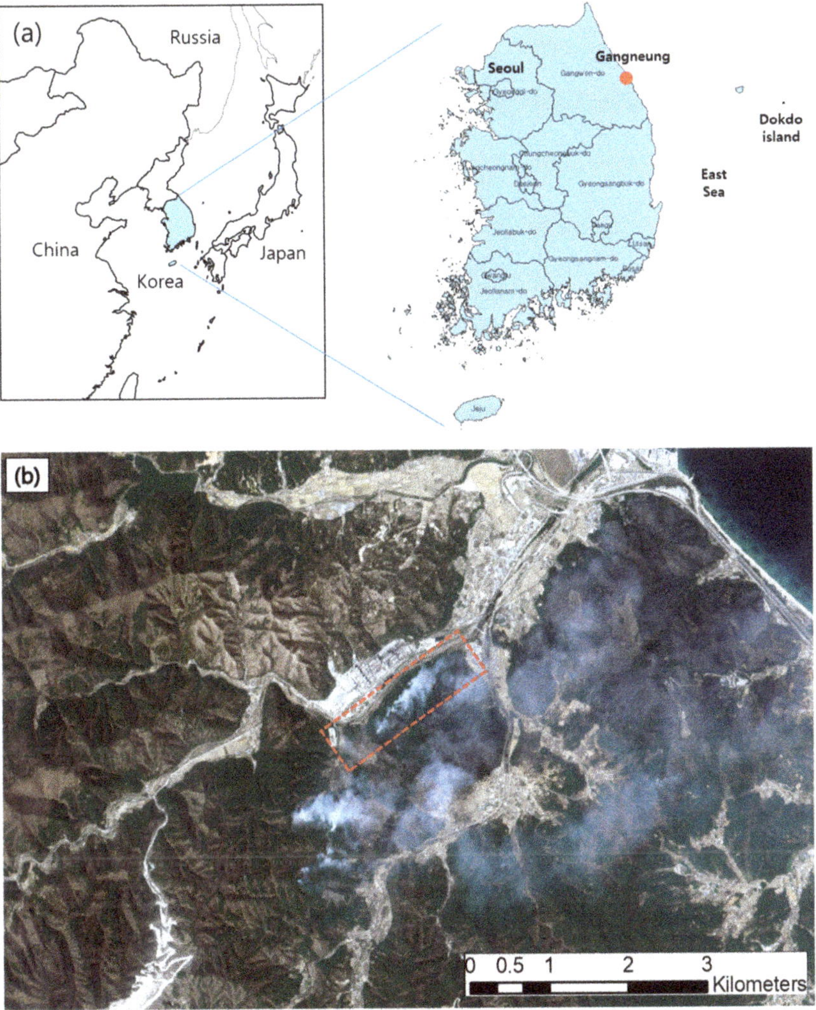

Figure 1. Location of (**a**) the city of Gangneung, and (**b**) the study area (red box) in the burning forest shown in a KOMPSAT-3A satellite image taken on 5 April 2019.

Figure 2 shows some of the damaged locations in the study area. The burned crown looks black along one ridge, and the boiled crown shows brown-colored needles in Figure 2a. In the burned surface area, green needles are distributed in the crown, although the ground surface and the bark of the lower trunks that burned are black (Figure 2b). A burned surface stresses the trees owing to a lack of organic matter and moisture in the soil.

Figure 2. The study area is forest damaged by fire from 4 to 5 April 2019, near the city of Gangneung: (**a**) burned forest where there are various burned (damaged) types mixed in the area; and (**b**) a typical burned surface area where some tree trunks were burned, even though the crown was not burned.

2.2. Data

2.2.1. Multispectral UAV Image

Multispectral images were acquired using a RedEdge camera (Micasense, Seattle, WA, USA) on 9 May 2019, which was more than one month after the forest fire. The camera was installed on a self-developed hexa-copter UAV that was 100 cm in diameter and 10 kg in weight. The acquired images were 214 scenes with 70% overlap and 50% sidelap. An image consisted of five bands (blue, green, red, red edge, and near infrared) that are appropriate to observe vegetation. The spatial resolution was 31 cm with a flight altitude of 500 m. The images were preprocessed to a mosaicked reflectance image using Pix4D software (Pix4 S.A., Prilly, Switzerland). Figure 3 shows an image with (a) natural color composite and (b) pseudo-infrared composite. The image shows some distinguishable colors in Figure 3a, which are dark brown, light brown, dark green, and light green for the burned crown, boiled crown, unburned pine trees, and deciduous trees, respectively. However, the burned surface and unburned (pine) trees cannot be classified from visual interpretation. Figure 3b shows more clearly distinguished colors using the NIR band. Figure 4 shows an NDVI transformed image.

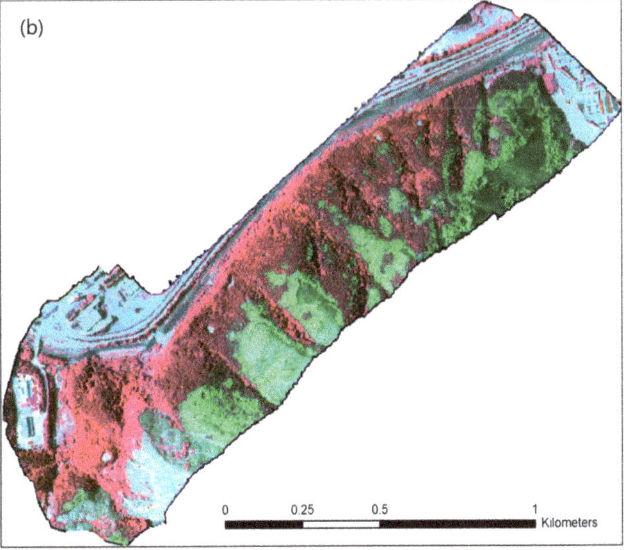

Figure 3. RedEdge multispectral unmanned aerial vehicle (UAV) image of some of the burned area (red box in Figure 1) near the city of Gangneung: (**a**) natural color composite: RGB = band 3, 2, 1; and (**b**) pseudo-infrared composite: RGB = band 5, 3, 2.

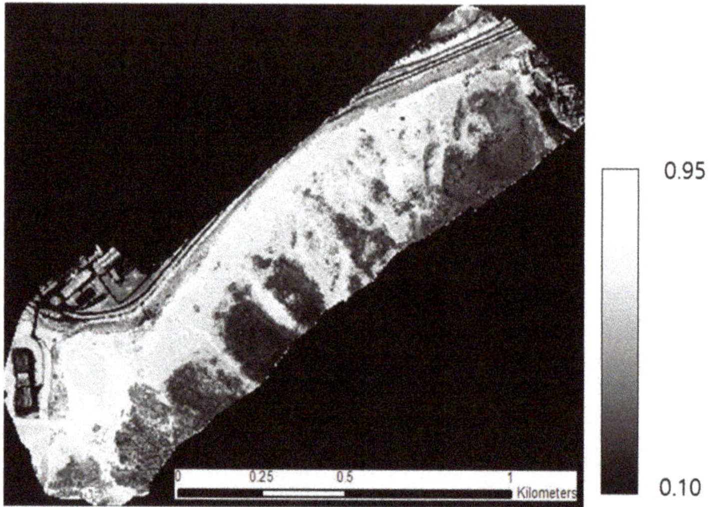

Figure 4. A normalized difference vegetation index (NDVI) transformed image that was stretched from 0.10 to 0.95.

2.2.2. Reference Map

A reference map was produced with screen digitizing based on a field survey, and it was used to extract samples to validate the classification results of burn severity. In screen digitizing, some classes were distinguished well by eye, such as burned crown, boiled crown, and unburned deciduous trees. The boundary between a burned surface and unburned pine was drawn on the image through a field survey. Figure 5 shows the reference map with color-coded classes. Red, orange, yellow, dark green, and light green mean burned crown, boiled crown, burned surface, unburned pine, and unburned deciduous, respectively. The classes are further defined in the following section.

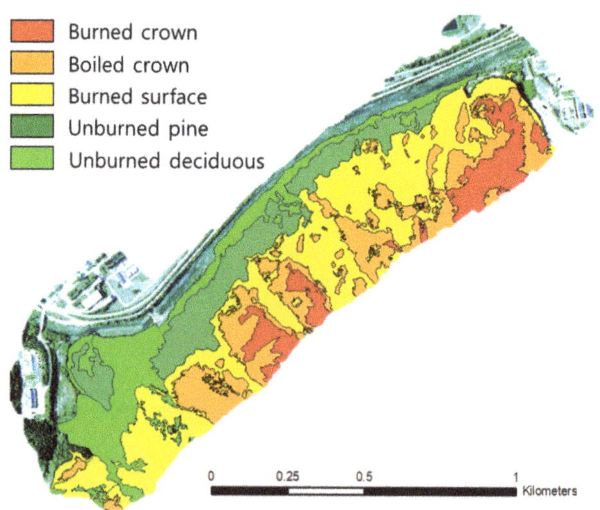

Figure 5. A reference map from visual interpretation based on a field survey.

3. Methods

3.1. Sample Collection and Spectral Analysis

3.1.1. Class Definition of Burn Severity

Previous studies defined classes of burn severity as extreme, high, moderate, low, and unburned. These classes might be subjective (or qualitative), and might not be considered spectral characteristics. In this study, burn severity classes are defined with consideration for both KCBI and spectral characteristics. The defined classes are burned crown, boiled crown, burned surface, unburned pine, and unburned deciduous in this study. Some classes can be compatible between KCBI and previously defined classes, such as burned crown (extreme), boiled crown (high), and burned surface (low). However, others are not compatible with each other. *Moderate* means a spatial mixture of burned crown and burned surface. *Unburned* should be classified as pine (coniferous) or deciduous, from the perspective of spectral characteristics.

3.1.2. Sample Collection

Samples were collected from a multispectral UAV image. The samples consisted of five defined classes as outlined in Section 3.1.1. Twenty plots were selected for each class based on a field survey and visual interpretation. The plots were evenly distributed in the image, and one plot is a nine-pixel square. Therefore, 450 sample pixels were collected for each class; 180 pixels (40%) were assigned as a training set, and the other 270 pixels (60%) were used as a validation set for the classifications. Figure 6 shows examples of the collected sample pixels.

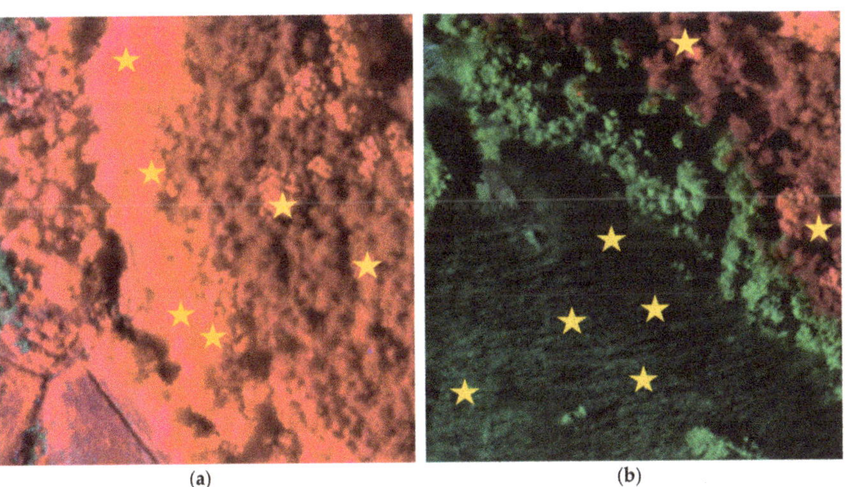

(a)   (b)

Figure 6. Examples of sample collections for (**a**) an unburned area and (**b**) a burned crown and burned surface area.

3.1.3. Spectral Characteristics Analysis

It is necessary to know the possible burn severity classifications as prior information. The possibilities were analyzed with statistics on reflectance and from spectral indices of the training samples. The mean and standard deviation of reflectance were calculated for each class. NDVI, red edge NDVI (RE-NDVI), and the visible-band difference vegetation index (VDVI) were calculated using the mean reflectance of each class. Equations (1) to (3) show the definitions of each vegetation index, where ρ is the mean reflectance of each band:

$$NDVI = \frac{\rho_{NIR} - \rho_{Red}}{\rho_{NIR} + \rho_{Red}},\tag{1}$$

$$RE - NDVI = \frac{\rho_{NIR} - \rho_{Red\ edge}}{\rho_{NIR} + \rho_{Red\ edge}},\tag{2}$$

$$VDVI = \frac{2 \times \rho_{Green} - \rho_{Red} - \rho_{Blue}}{2 \times \rho_{Green} + \rho_{Red} + \rho_{Blue}}.\tag{3}$$

Table 1 shows the mean values of the burn severity classes for three spectral indices where mean values were estimated from collected training samples. NDVI shows bigger gaps among the classes than other indices. NDVI might be useful in the classification of burn severity because we can easily define thresholds among classes.

Table 1. Mean of burn severity classes for each vegetation index. NDVI, normalized difference vegetation index; RE-NDVI, red edge NDVI; VDVI, visible-band difference vegetation index.

Index	Burned Crown	Boiled Crown	Burned Surface	Unburned Pine	Unburned Deciduous
NDVI	0.33	0.31	0.78	0.84	0.91
RE-NDVI	0.33	0.27	0.45	0.49	0.51
VDVI	0.08	0.00	0.33	0.36	0.52

3.2. Classification of Burn Severity

3.2.1. Supervised Classification

Maximum likelihood (MLH) and spectral angle mapper (SAM) were used as supervised classification methods. The MLH classifier assigns a pixel to a class with the highest probability under the assumption that reflectance values of each class have a normal (Gaussian) distribution in each band. The probability for a pixel is calculated by the multivariate normal density function from the mean, variance, and covariance of training samples [27]. The SAM classifier calculates similarity using the spectral angle between a pixel and the mean of each class. Spectral reflectance is assumed to be a vector in n-dimensional space, where n is the number of bands. A pixel is assigned to a class with the smallest spectral angle [28]. Equations (4) and (5) show the definitions of MLH and SAM, respectively:

$$P(X|w_i) = \frac{1}{(2\pi)^{\frac{n}{2}}|V_i|^{\frac{1}{2}}} \exp[-\frac{1}{2}(X - M_i)^T V_i^{-1}(X - M_i),\tag{4}$$

where, n, X, V_i, and M_i denote the number of multispectral bands, the unknown measurement vector, the covariance matrix of each training class, and the mean vector of each training class, respectively, and

$$\alpha = cos^{-1}\left(\frac{\sum_{i=1}^{n} t_i r_i}{\left(\sum_{i=1}^{n} t_i^2\right)^{\frac{1}{2}}\left(\sum_{i=1}^{n} r_i^2\right)^{\frac{1}{2}}}\right),\tag{5}$$

where, n, t_i, and r_i denote the number of multispectral bands, the unknown measurement vector, and the reference spectrum vector, respectively; and α denotes an angle between r and t vectors.

3.2.2. Spectral Index Classification

Thresholding of a spectral index is used a classification method for burn severity. In this study, NDVI was used only as a spectral index for classification because NDVI shows higher differences among classes, compared with other indices. Figure 7 shows the range of NDVI values for each class, with the mean ± standard deviation. Burned crown and boiled crown are perfectly separated with burned surface, unburned pine, and unburned deciduous. However, some class pairs overlap each other, such as burned crown (boiled crown) and burned surface (unburned pine). Thresholds were defined as the median of the overlapping range between neighboring classes (Table 2).

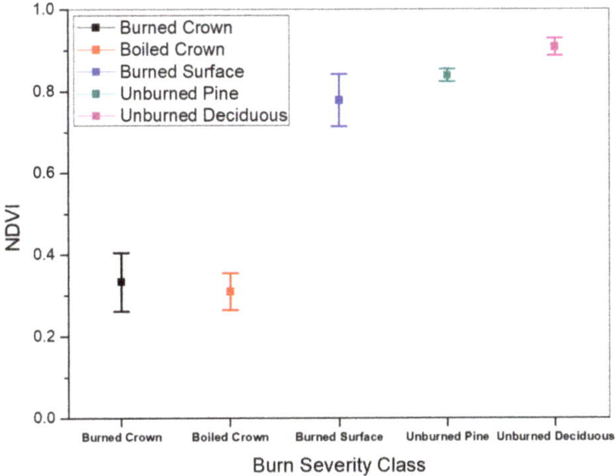

Figure 7. Distribution of NDVI values with mean ± one standard deviation for each burn severity class.

Table 2. Thresholds between burn severity classes using the median of the overlapped range.

Class Pair	Burned Crown—Boiled Crown	Boiled Crown—Burned Surface	Burned Surface —Unburned Pine	Unburned Pine—Unburned Deciduous
Threshold	0.308	0.565	0.833	0.870

4. Results

4.1. Spectral Charateristics

Spectral reflectance curves were plotted for training samples; those are the mean reflectance of each band (Figure 8). Burned crown shows lower reflectance than other classes because of soot and ash. In boiled crown, yellow needles show high reflectance in green (560 nm) and red (668 nm) bands, and needles are distributed in the crown, which creates higher reflectance in the near-infrared band (840 nm) than burned crown, where we can see that the spectral reflectance of boiled crown is higher in the blue (475 nm) and red bands than with unburned pine, unburned deciduous, and burned surface classes. This can be attributed to the loss of chlorophyll, a pigment that absorbs blue and red as the leaves turn yellow owing to damage from the heat. Unburned deciduous shows low reflectance in the blue and red bands, and high reflectance in the green, red edge, and NIR bands; these are typical spectral characteristics of broadleaf vegetation. This study focuses on unburned pine and burned surface classifications. Burned surface and unburned pine show similar spectral reflectance curves and an overlapped range for the mean ± standard deviation (Figure 9), because they both have green needles at the top of the crown. However, a burned surface shows slightly higher reflectance at the red edge, and lower reflectance at the NIR bands. This is called the red-edge shift phenomenon, where the center wavelength of the red edge between the red and NIR bands slightly shifts to a shorter wavelength [29]. The reason for the red-edge shift is known to be decreased vitality owing to stress. Therefore, a burned surface and unburned pine might be classified using slightly different spectral characteristics.

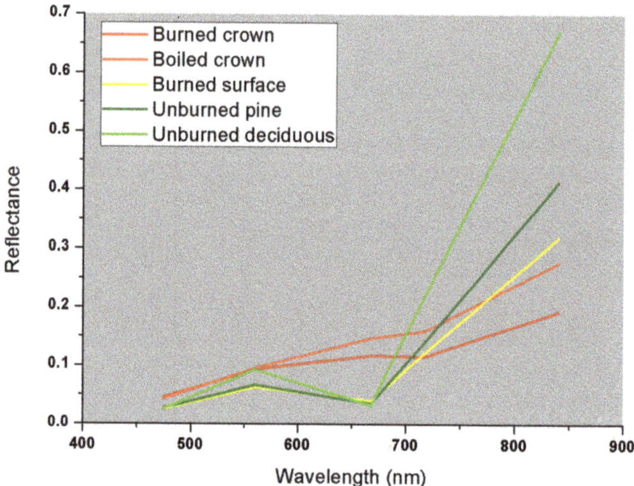

Figure 8. Mean spectral reflectance curve for five classes in the training samples.

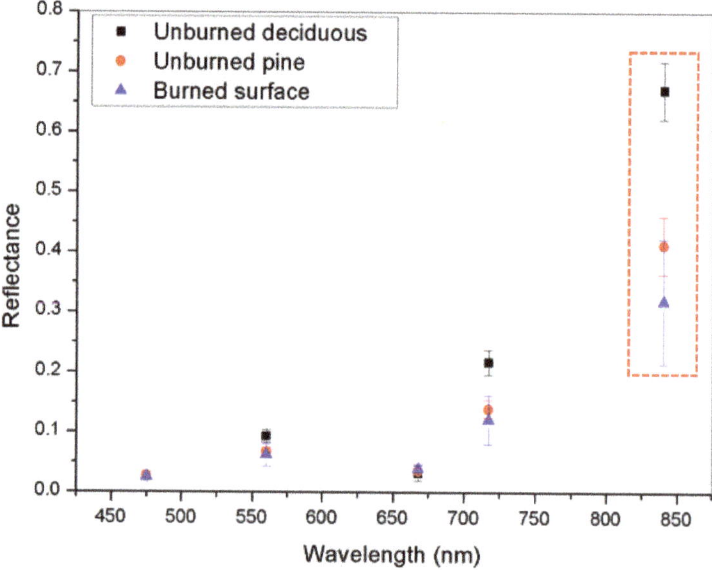

Figure 9. Range of mean ± one standard deviation for burned surface, unburned pine, and unburned deciduous.

4.2. Interpretation of Classification Results

Burn severity was classified using two supervised classification methods plus the NDVI thresholding method. Figure 10a–c show classification results from using MLH, SAM, and NDVI thresholding, respectively. SAM shows good classification results for overall classes, except for burned surface, which was mostly misclassified as unburned pine. MLH showed better results with a burned surface, although it was overestimated by misclassification between unburned pine and burned surface. Additionally, it showed confusion between unburned pine and unburned deciduous. NDVI thresholding showed moderate results among the three classification methods. The burned surface class was underestimated, and burned crown was confused with boiled crown.

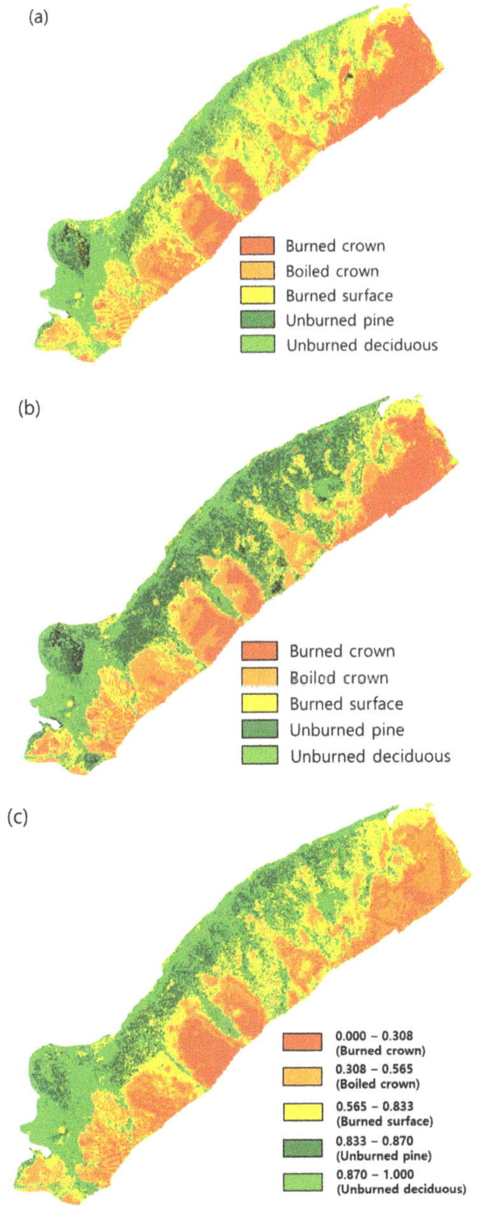

Figure 10. Results from classification of burn severity using (**a**) maximum likelihood (MLH), (**b**) spectral angle mapper (SAM), and (**c**) NDVI thresholding.

Confusion matrices (Tables 3–5) show the classification accuracy for each classifier. High accuracy was shown in the following order: MLH, SAM, and NDVI thresholding. Overall accuracies are 89%, 81%, and 71% for MLH, SAM, and NDVI thresholding, respectively. Kappa coefficients were also similar to overall accuracy at 0.86, 0.76, and 0.64, respectively. As seen in Table 3, MLH showed very high accuracy of more than 85% for four classes, except unburned pine. We can see that 33% of the unburned pine was misclassified as burned surface, demonstrating that burned surface was overestimated, as seen in the visual analysis of classification results in Figure 10a. In Table 4, we see that SAM had more than 85% classification accuracy in three classes except for unburned pine and unburned deciduous. Those were misclassified as burned surface, or confused with each other, because the SAM algorithm uses a pattern of spectral reflectance rather than an absolute value of reflectance. In Table 5, NDVI thresholding showed lower overall accuracy than the supervised classifiers, MLH and SAM. Low overall accuracy was caused by confusion between burned crown and boiled crown from similar NDVI values for burned out or discolored needles. However, the classification accuracy of burned surface, unburned pine, and unburned deciduous showed similar or higher levels than the two supervised classifiers.

Table 3. Confusion matrix for classification accuracy assessment using MLH (%).

Reference Classifications	Burned Crown	Boiled Crown	Burned Surface	Unburned Pine	Unburned Deciduous	Total
Burned crown	96	13	0	0	0	22
Boiled crown	4	87	3	0	0	19
Burned surface	0	0	92	33	0	25
Unburned pine	0	0	5	67	0	14
Unburned deciduous	0	0	0	0	100	20
Total	100	100	100	100	100	100

Overall accuracy = 89%, Kappa coefficient = 0.86

Table 4. Confusion matrix for classification accuracy assessment using spectral angle mapper (SAM) (%).

Reference Classifications	Burned Crown	Boiled Crown	Burned Surface	Unburned Pine	Unburned Deciduous	Total
Burned crown	95	6	0	0	0	20
Boiled crown	2	91	2	0	0	19
Burned surface	3	3	85	24	20	27
Unburned pine	0	0	13	74	22	22
Unburned deciduous	0	0	0	2	58	12
Total	100	100	100	100	100	100

Overall accuracy = 81%, Kappa coefficient = 0.76

Table 5. Confusion matrix for classification accuracy assessment using NDVI thresholding (%).

Reference Classifications	Burned Crown	Boiled Crown	Burned Surface	Unburned Pine	Unburned Deciduous	Total
Burned crown	72	70	6	0	0	29
Boiled crown	25	27	0	0	0	11
Burned surface	3	3	89	24	0	24
Unburned pine	0	0	5	74	6	17
Unburned deciduous	0	0	0	2	94	19
Total	100	100	100	100	0	100

Overall accuracy = 71%, Kappa coefficient = 0.64

5. Discussion

In this study, two supervised classifiers and NDVI thresholding were compared for the classification of forest burn severity, including burned surface. As a result, the classification of burned surface showed accuracy of more than 85%. This means that the UAV multispectral image can be used to accurately classify a burned surface, even though the two classes are very similar when viewed from the air.

Comparing the classification methods, the supervised classifiers (MLH and SAM) showed more than 80% overall accuracy, while the NDVI thresholding accuracy was 71%. However, 20% to 30% of unburned pine and deciduous trees were misclassified as a burned surface, and the amount of burned surface was overestimated. This means that a high degree of expertise and a lot of time are required to collect dozens to hundreds of training samples for a supervised classifier such as MLH, while the classification accuracy between burned surface and unburned pine is not high. SAM can classify with one training sample, such as a collected pixel or a spectrum from a spectral library. It showed a high number of misclassifications between classes with similar spectral patterns, although hundreds of training samples were used.

NDVI thresholding showed similar or higher accuracy with the burned surface class, compared with the supervised classifiers, even though it showed confusion between burned crown and boiled crown. Previous studies reported that burned crown and boiled crown can be classified easily with various methods, including visual interpretation. Detection or classification of a burned surface should be the focus for accurate assessment of damaged areas and burn severity. Therefore, the NDVI thresholding method is expected to be able to estimate forest fire damage more easily and accurately. Further studies are needed to generalize the threshold for application in the field. In other words, independent thresholds should be defined for different regions and times when classifying burn severity of a forest.

UAV multispectral images have very high spatial resolution and multispectral bands that can be used as training samples for classification using high-resolution satellite images and deep learning algorithms in the future. Deep learning can enhance accuracy and convenience, although it needs a long-term approach to collecting a large number of training samples. High-resolution earth observation satellites can also be a useful tool for analyzing burn severity and damaged areas. The Korean government is developing a new earth observation satellite program, the Compact Advanced Satellite 500 (CAS 500), to shorten the observation interval. The first and second twin satellites will be launched in 2020, and have a 0.5 m panchromatic band and 2 m multispectral bands. We are expecting acquisition of high-resolution multispectral images within two to three days. In the future, it will be necessary to suggest the possibility of classifying burn severity using high-resolution satellite images, in comparison with UAV images.

6. Conclusions

This study tried to analyze the use of multispectral UAV images to classify burn severity, including burned surfaces. A RedEdge multispectral image was acquired for a region of the Gangneung forest fire in April 2019. Spectral characteristics showed differences among burn severity classes, although some of them were similar. Burn severity was classified using two supervised classifiers, MLH and SAM, as well as the NDVI thresholding method. Classification accuracies were about 80% to 90% using the supervised classifiers and about 70% using NDVI thresholding. They showed an accuracy of more than 85% for the burned surface class, where a multispectral UAV image can differentiate a burned surface from unburned pine or deciduous trees. The NDVI thresholding method also showed high classification accuracy for burned surface, unburned pine, and deciduous. It can be useful as an easier and more accurate tool for the estimation of burn severity and damaged areas than a supervised classifier. Supervised classification approaches might be applied to other regions through collection of corresponding training samples. However, NDVI of burn severity classes might have different values by regional characteristics. Further studies are needed to generalize NDVI or the thresholds for application in other regions. In the future, multispectral UAV images can also be used for training deep learning techniques and high-resolution satellite images.

Author Contributions: Conceptualization, C.-s.W. and J.-i.S.; data curation, C.-s.W.; formal analysis, J.-i.S. and W.-w.S.; funding acquisition, J.P. and T.K.; investigation, J.-i.S., C.-s.W., and W.-w.S.; methodology, J.-i.S.; project administration, J.P. and T.K.; supervision, C.-s.W.; visualization, J.-i.S. and W.-w.S.; writing—original draft, J.-i.S. and W.-w.S.; writing—review and editing, J.-i.S., W.-w.S. and T.K.

Funding: This research was funded by "National Institute of Forest Science grant, number FE0500-2018-04" and "Satellite Information Utilization Center Establishment Program (18SIUE-B148326-01) by Ministry of Land, Infrastructure, and Transport of Republic of Korea".

Acknowledgments: We thank In-Sup Kim as the staff of the National Institute of Forest Science who helped our UAV image acquisition and we also thank Korea Aerospace Research Institute for offering a KOMPSAT-3A satellite image.

Conflicts of Interest: The authors declare they have no conflicts of interest. The funders had no role in the design of the study; in the collection, analyses, or interpretation of the data; in the writing of the manuscript; or in the decision to publish the results.

References

1. Statistical yearbook of wild fire 2018, Statistics Korea. Available online: https://www.index.go.kr/potal/stts/idxMain/selectPoSttsIdxMainPrint.do?idx_cd=1309&board_cd=INDX_001 (accessed on 16 September 2019).
2. Tang, L.; Shao, G. Drone remote sensing for forestry research and practices. *J. For. Res.* **2015**, *26*, 791–797. [CrossRef]
3. Torresan, C.; Berton, A.; Carotenuto, F.; Di Gennaro, S.F.; Gioli, B.; Matese, A.; Wallace, L. Forestry applications of UAVs in Europe: A review. *Int. J. Remote Sens.* **2017**, *38*, 2427–2447. [CrossRef]
4. Sankey, T.; Donager, J.; McVay, J.; Sankey, J.B. UAV lidar and hyperspectral fusion for forest monitoring in the southwestern USA. *Remote Sens. Environ.* **2017**, *195*, 30–43. [CrossRef]
5. Carvajal-Ramírez, F.; Marques da Silva, J.R.; Agüera-Vega, F.; Martínez-Carricondo, P.; Serrano, J.; Moral, F.J. Evaluation of Fire Severity Indices Based on Pre-and Post-Fire Multispectral Imagery Sensed from UAV. *Remote Sens.* **2019**, *11*, 993. [CrossRef]
6. Pla, M.; Duane, A.; Brotons, L. Potential of UAV images as ground-truth data for burn severity classification of Landsat imagery: Approaches to an useful product for post-fire management. *Rev. Teledetec.* **2017**, *49*, 91–102. [CrossRef]
7. Fernández-Guisuraga, J.; Sanz-Ablanedo, E.; Suárez-Seoane, S.; Calvo, L. Using unmanned aerial vehicles in postfire vegetation survey campaigns through large and heterogeneous areas: Opportunities and challenges. *Sensors* **2018**, *18*, 586. [CrossRef]
8. García, M.L.; Caselles, V. Mapping burns and natural reforestation using Thematic Mapper data. *Geocarto Int.* **1991**, *6*, 31–37. [CrossRef]
9. Barbosa, P.M.; Stroppiana, D.; Grégoire, J.M.; Cardoso Pereira, J.M. An assessment of vegetation fire in Africa (1981–1991): Burned areas, burned biomass, and atmospheric emissions. *Glob. Biogeochem. Cycles* **1999**, *13*, 933–950. [CrossRef]
10. Chafer, C.J.; Noonan, M.; Macnaught, E. The post-fire measurement of fire severity and intensity in the Christmas 2001 Sydney wildfires. *Int. J. Wildland Fire* **2004**, *13*, 227–240. [CrossRef]
11. Epting, J.; Verbyla, D.; Sorbel, B. Evaluation of remotely sensed indices for assessing burn severity in interior Alaska using Landsat TM and ETM+. *Remote Sens. Environ.* **2005**, *96*, 328–339. [CrossRef]
12. Collins, B.M.; Kelly, M.; van Wagtendonk, J.W.; Stephens, S.L. Spatial patterns of large natural fires in Sierra Nevada wilderness areas. *Landsc. Ecol.* **2007**, *22*, 545–557. [CrossRef]
13. Tran, B.; Tanase, M.; Bennett, L.; Aponte, C. Evaluation of Spectral Indices for Assessing Fire Severity in Australian Temperate Forests. *Remote Sens.* **2018**, *10*, 1680. [CrossRef]
14. Amos, C.; Petropoulos, G.P.; Ferentinos, K.P. Determining the use of Sentinel-2A MSI for wildfire burning & severity detection. *Int. J. Remote Sens.* **2019**, *40*, 905–930.
15. Filipponi, F. BAIS2: Burned Area Index for Sentinel-2. *Proc. Int. Electron. Conf. Remote Sens.* **2018**, *2*, 364. [CrossRef]

16. Chuvieco, E.; Martin, M.P.; Palacios, A. Assessment of different spectral indices in the red-near-infrared spectral domain for burned land discrimination. *Int. J. Remote Sens.* **2002**, *23*, 5103–5110. [CrossRef]

17. Key, C.H.; Benson, N.C. *Fire Effects Monitoring and Inventory Protocol—Landscape Assessment*; USDA Forest Service Fire Science Laboratory: Missoula, MT, USA, 2002; pp. 1–16.

18. Van Wagtendonk, J.W.; Root, R.R.; Key, C.H. Comparison of AVIRIS and Landsat ETM+ detection capabilities for burn severity. *Remote Sens. Environ.* **2004**, *92*, 397–408. [CrossRef]

19. Hartford, R.A.; Frandsen, W.H. When it's hot, it's hot... or maybe it's not! (Surface flaming may not portend extensive soil heating). *Int. J. Wildland Fire* **1992**, *2*, 139–144. [CrossRef]

20. Neary, D.G.; Klopatek, C.C.; DeBano, L.F.; Ffolliott, P.F. Fire effects on belowground sustainability: A review and synthesis. *For. Ecol. Manag.* **1999**, *122*, 51–71. [CrossRef]

21. Miller, J.D.; Yool, S.R. Mapping forest post-fire canopy consumption in several overstory types using multi-temporal Landsat TM and ETM data. *Remote Sens. Environ.* **2002**, *82*, 481–496. [CrossRef]

22. Smith, A.M.; Wooster, M.J. Remote classification of head and backfire types from MODIS fire radiative power and smoke plume observations. *Int. J. Wildland Fire* **2005**, *14*, 249–254. [CrossRef]

23. McKenna, P.; Erskine, P.D.; Lechner, A.M.; Phinn, S. Measuring fire severity using UAV imagery in semi-arid central Queensland, Australia. *Int. J. Remote Sens.* **2017**, *38*, 4244–4264. [CrossRef]

24. Fraser, R.; Van Der Sluijs, J.; Hall, R. Calibrating satellite-based indices of burn severity from UAV-derived metrics of a burned boreal forest in NWT, Canada. *Remote Sens.* **2017**, *9*, 279. [CrossRef]

25. Rossi, F.; Fritz, A.; Becker, G. Combining satellite and UAV imagery to delineate forest cover and basal area after mixed-severity fires. *Sustainability* **2018**, *10*, 2227. [CrossRef]

26. Won, M. Analysis of Burn Severity in Large-Fire Area Using Satellite Imagery. Ph.D. Thesis, Korea University, Seoul, Korea, July 2013.

27. Swain, P.H.; Davis, S.M. Remote sensing: The quantitative approach. *IEEE Trans. Pattern Anal. Mach. Intell.* **1981**, *6*, 713–714. [CrossRef]

28. Kruse, F.A.; Lefkoff, A.B.; Boardman, J.W.; Heidebrecht, K.B.; Shapiro, A.T.; Barloon, P.J.; Goetz, A.F.H. The spectral image processing system (SIPS)—Interactive visualization and analysis of imaging spectrometer data. *Remote Sens. Environ.* **1993**, *44*, 145–163. [CrossRef]

29. Clevers, J.P.G.W.; Jongschaap, R. Imaging spectrometry for agricultural applications. In *Imaging Spectrometry—Basic Priciples and Prospective Applications*, 1st ed.; Van Der Meer, F.D., Ed.; Kluwer Academic Publishers: Dordrecht, The Netherlands, 2003; pp. 157–199.

Article

Characterizing Seedling Stands Using Leaf-Off and Leaf-On Photogrammetric Point Clouds and Hyperspectral Imagery Acquired from Unmanned Aerial Vehicle

Mohammad Imangholiloo [1,2,*], Ninni Saarinen [1,2,3], Lauri Markelin [4], Tomi Rosnell [4], Roope Näsi [4], Teemu Hakala [4], Eija Honkavaara [4], Markus Holopainen [1,2], Juha Hyyppä [2,4] and Mikko Vastaranta [2,3]

[1] Department of Forest Sciences, University of Helsinki, P. O. Box 27, 00014 Helsinki, Finland; Ninni.Saarinen@helsinki.fi (N.S.); Markus.Holopainen@helsinki.fi (M.H.)
[2] Centre of Excellence in Laser Scanning Research, Finnish Geospatial Research Institute, National Land Survey, Geodeetinrinne 2, 02431 Masala, Finland; Juha.Hyyppa@nls.fi (J.H.); Mikko.Vastaranta@uef.fi (M.V.)
[3] School of Forest Sciences, University of Eastern Finland, P. O. Box 111, 80101 Joensuu, Finland
[4] Department of Remote Sensing and Photogrammetry, Finnish Geospatial Research Institute (FGI), National Land Survey of Finland (NLS), Geodeetinrinne 2, 02430 Masala, Finland; Lauri.Markelin@nls.fi (L.M.); Tomi.Rosnell@nls.fi (T.R.); Roope.Nasi@nls.fi (R.N.); Teemu.Hakala@nls.fi (T.H.); Eija.Honkavaara@nls.fi (E.H.)
* Correspondence: Mohammad.Imangholiloo@helsinki.fi; Tel.: +358-29-415-8181

Received: 24 April 2019; Accepted: 8 May 2019; Published: 13 May 2019

Abstract: Seedling stands are mainly inventoried through field measurements, which are typically laborious, expensive and time-consuming due to high tree density and small tree size. In addition, operationally used sparse density airborne laser scanning (ALS) and aerial imagery data are not sufficiently accurate for inventorying seedling stands. The use of unmanned aerial vehicles (UAVs) for forestry applications is currently in high attention and in the midst of quick development and this technology could be used to make seedling stand management more efficient. This study was designed to investigate the use of UAV-based photogrammetric point clouds and hyperspectral imagery for characterizing seedling stands in leaf-off and leaf-on conditions. The focus was in retrieving tree density and the height in young seedling stands in the southern boreal forests of Finland. After creating the canopy height model from photogrammetric point clouds using national digital terrain model based on ALS, the watershed segmentation method was applied to delineate the tree canopy boundary at individual tree level. The segments were then used to extract tree heights and spectral information. Optimal bands for calculating vegetation indices were analysed and used for species classification using the random forest method. Tree density and the mean tree height of the total and spruce trees were then estimated at the plot level. The overall tree density was underestimated by 17.5% and 20.2% in leaf-off and leaf-on conditions with the relative root mean square error (relative RMSE) of 33.5% and 26.8%, respectively. Mean tree height was underestimated by 20.8% and 7.4% (relative RMSE of 23.0% and 11.5%, and RMSE of 0.57 m and 0.29 m) in leaf-off and leaf-on conditions, respectively. The leaf-on data outperformed the leaf-off data in the estimations. The results showed that UAV imagery hold potential for reliably characterizing seedling stands and to be used to supplement or replace the laborious field inventory methods.

Keywords: seedling stand inventorying; photogrammetric point clouds; hyperspectral imagery; unmanned aerial vehicles; leaf-off; leaf-on

1. Introduction

Sustainable forest management requires accurate and up-to-date information. The information is acquired by field measurements or remote sensing-based inventorying. The field measurements are time-consuming, expensive and laborious, in contrast to remote sensing-based inventorying techniques. Currently, unmanned aerial vehicles (UAVs, aka: drones) are in high interest for forest inventorying because of the UAVs' capability to collect data from which suite of essential forest inventory attributes can be derived with accuracy close to field inventories [1–8]. UAVs provide an easy, inexpensive, and repeatable data collection method [9] with very high spatial resolution data that can even support the detection of small trees which has not been possible using airborne laser scanning (ALS) data [10].

In Finland, seedling stands are defined as the forest stands with mean height of < 7 m (conifer) or 9 m (deciduous) [11]. Conditions of the seedling stands can greatly predict and define the condition of future mature stands [12]. For example, Huuskonen and Hynynen [12] revealed that precommercial thinning, which was carried out when the dominate height was 3 m and the target tree density was 2000 trees per hectare (TPH), resulted in an increase of 15% in the mean diameter of the first commercial thinning. Thus, monitoring and management of the seedling stands development are required to ensure quality timber as well as the future timber supply.

ALS data, used in operational private forest inventories (61%) in Finland, is not capable to characterise seedling stands due to small tree size and high tree density. Thus, the seedling stands are inventoried by field measurements, which are the most expensive part of the total cost of the inventory. Therefore, analysing other cost-efficient means to estimate tree density, height, and species composition is required. A few studies explored the use of UAV-based photogrammetric point clouds in the seedling stands. For example, Puliti et al. [13] estimated biophysical properties of seedling stands using UAV-photogrammetric point cloud data, and compared it with ALS data. Similarly, UAV demonstrated promising results to detect coniferous seedlings in leaf-off conditions where seedlings were visually and spectrally distinctive [14]. Moreover, Goodbody et al. [15] combined UAV- and aerial-photogrammetric point clouds to assess spatial, spectral and structural details for the seedling stands. The UAV-based photogrammetric data were also utilized to investigate the feasibly and merits of UAV for evaluating regeneration performance in naturally-growing and planted conifer seedlings in different growth phases [16]; as well as assessing the effects of the European spruce bark beetle (*Ips typographus* L.) disturbance on natural regeneration and standing deadwood [17].

In addition to the few UAV studies, other remote sensing materials were also used for investigating the seedling stands, for example airborne imagery [18–21] and SPOT-5 satellite imagery [22]. Moreover, ALS data were applied for analysing small trees in the forest-tundra ecotone [23,24], regeneration or young forests [25,26] and predicting aboveground biomass (AGB) change in young forests [27]. Additionally, Korpela et al. [28] combined ALS and airborne imagery for characterizing seedling stand vegetation; and for detecting the requirement for tending seedling stands [29]. Also, Korhonen et al. [29] used ALS and aerial imagery to detect the tending requirement of seedling stands, by creating model function based on ALS-derived echo intensity and height percentiles together with aerial images texture. However, they appointed out that their approach could not completely replace the field visits with regards to the need for tending seedling stands.

The use of hyperspectral data and comparing data from leaf-off and leaf-on conditions remained unexplored in the previous seedling-focused studies. Thus, this study was designed to extend current knowledge of using UAV-red, green, blue (RGB)-imagery, UAV-hyperspectral data as well as analysing the performance of leaf-off and leaf-on data with predefined plot-level tree densities (TPH). This research concentrates on retrieving the total and spruce-specific tree density and height in seedling stands in the southern boreal forests of Finland. We focused on the spruces because it is the species of interest to be grown and spruce seedling stands commonly require more care (e.g., tending and removal of naturally regenerated broadleaf trees) compared to seedling stands of Scots pine. In Finland, seedling stands are divided into young (height ≤ 1.3 m, YoS) and advanced seedling stands (height

> 1.3 m, AdS). Therefore, this study also aims to assess the differences between predictions for YoS and AdS.

2. Materials and Methods

2.1. Study Area and Establishment of the Sample Plots

This study was carried out in a southern boreal forest zone in Evo, Finland (61.20° N, 25.08° E, 133–150 m above sea level) (Figure 1). There are mostly managed forests where Scots pine and Norway spruce are the dominant tree species.

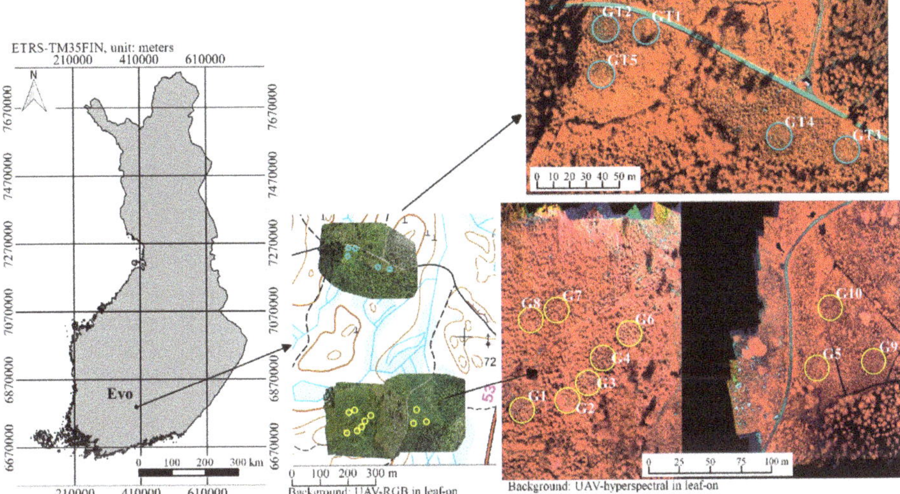

Figure 1. Map of study area and established sample plots in young seedling stand (northern image block) and in advanced seedling stand (southern image blocks). Background: UAV-red, green, blue (RGB)-image mosaics (middle map) and UAV-hyperspectral image mosaics (right maps) visualized coloured-infrared (885.9 nm, 605.4 nm, 513.5 nm) in leaf-on conditions.

In our study, we selected one YoS and one AdS stand from the study area based on the existing forest resource information. A prerequisite for both seedling stands was the number of TPH, which had to be more than 2400. Such a density was required to establish sample plots and thin them to varying densities. We established five sample plots with an 8-m radius to YoS and 10 sample plots with a 10-m radius to AdS. The sample plots in YoS were thinned approximately to the following target tree densities: 1200, 1400, 1600, 1800, and 2000 TPH. Respectively, the sample plots in AdS were thinned to the following densities: 600, 800, 1000, 1200, 1400, 1600, 1800, 2000, 2200, and 2400 TPH.

The sample plots were established in April through May 2016. Sample plot locations were recorded using the Trimble GeoXT Global Navigation Satellite System (GNSS) device. The GNSS positions were differentially corrected using the data from the local reference station. The expected accuracy in an open area is below 1 meter. After the thinning treatments, tree attributes were measured during June (Table 1), and the sample plot-level forest inventory attributes were compiled. In YoS all remaining trees were spruce, but AdS sample plots had an admixture of birch that varied from 0% to 51%. The site type of our sample plots is the mesic heath forest.

Table 1. Tree height variation in sample plots. YoS: young seedling stands, AdS: advanced seedling stands, H_{min}, H_{max}, H_{mean}, and H_{std}: minimum, maximum, mean and standard deviation of field-measured heights (TPH: Trees per hectare, Height unit: meter).

Stand Development Class	Plot Name	Total Trees			Spruce					Birch				
		Stem Number (TPH)	H_{mean} (m)	Stem Number	H_{min}	H_{max}	H_{mean}	H_{std}	Stem Number	H_{min}	H_{max}	H_{mean}	H_{std}
YoS ($n = 5$)	GT1	1989	1.19	1989	0.73	1.87	1.19	0.33	0				
	GT2	1790	1.16	1790	0.77	1.78	1.16	0.24	0				
	GT3	1194	1.12	1194	0.77	1.61	1.12	0.19	0				
	GT4	1393	1.05	1393	0.82	1.48	1.05	0.20	0				
	GT5	1592	1.14	1592	0.86	1.56	1.14	0.19	0				
AdS ($n = 10$)	G1	986	2.78	891	1.62	3.92	2.66	0.55	95	3.66	4.33	3.90	0.4
	G2	605	3.23	446	1.87	3.83	3.00	0.57	159	3.36	4.27	3.88	0.4
	G3	1592	3.20	1369	1.71	4.00	3.12	0.58	223	3.28	4.33	3.67	0.3
	G4	1814	3.66	1273	1.57	4.40	3.44	0.63	541	3.36	5.01	4.17	0.5
	G5	1401	2.70	1401	1.62	3.87	2.70	0.54	0				
	G6	2069	3.08	2037	1.71	4.54	3.07	0.66	32	3.44	3.44	3.44	
	G7	2228	3.72	1464	2.09	4.21	3.27	0.52	764	3.36	5.58	4.58	0.6
	G8	2388	3.74	1178	1.62	4.28	3.35	0.62	1210	2.74	5.17	4.12	0.6
	G9	1210	2.45	1210	1.71	3.56	2.45	0.50	0				
	G10	796	2.95	796	2.03	3.96	2.95	0.58	0				

During the field data collection in June, many fern and grasses had emerged. In AdS, the tree species was determined, and the diameter at breast height (DBH) was measured with steel callipers from every tree with a height of ≥ 1.3 m. The tree height was measured using an electronic hypsometer (Vertex IV, Haglöf, Långsele, Sweden) from every third tree for each tree species. In addition, the height of the tallest tree was measured from every sample plot. YoS was measured similarly, but instead of DBH, the diameter at ground height was measured because the mean height of the YoS sample plots was less than 1.3 m (Table 1). The height for all of the trees within each sample plot was predicted using the sample tree height measurements and by fitting the Näslund's height curve [30] to the measured data. The relative RMSE of the tree height prediction was 12.8% (relative Bias: −0.13%) and 11.8% (relative Bias: 0.60%) for YoS and AdS, respectively. In the sample plots with mixed species classes, mean tree height of all trees was calculated with a weighted average of the number of each tree species and their mean height. Plot-level TPH was calculated by dividing the number of field-observed trees in each plot with its area (radius of 8 and 10 m in YoS and AdS, respectively) and converting the area to hectare. Species-specific tree height and TPH statistics are presented in Table 1.

2.2. Remote Sensing Data

Remote sensing data acquisition were carried out using a hexacopter drone of the Finnish Geospatial Research Institute (FGI). A hyperspectral camera based on Fabry–Pérot interferometer (FPI) [31] and a Samsung NX300 RGB camera were used to collect remote sensing imagery. The FPI technology provides spectral data cubes with a rectangular image format, but each band in the data cube has a slightly different position and orientation. The sensor provides images with dimension of 1024 in 648 pixels where every pixel is 11 μm × 11 μm. In this study, a filter with a wavelength range of 500–900 nm and settings with 36 separate bands was used; the spectral resolution range was 10–40 nm at the full width at half maximum (FWHM) (Table 2). A Samsung RGB camera had a 16-mm fixed lens and an image size of 5472 × 3648 pixels. The drone was equipped with a NV08C-CSM-GNSS receiver that was used to calculate the flight trajectory. The Raspberry Pi2 on-board computer was used for collecting timing data for all devices and for logging the GNSS receiver. More details of the imaging sensor and UAV system are provided in [32,33].

Table 2. Spectral settings of the hyperspectral spectral camera. L0: Central wavelength, FWHM: Full width at half maximum.

Spectral Settings of the Hyperspectral Spectral Camera											
L0 (nm)	507.24	509.08	513.48	520.44	537.16	545.62	554.2	562.85	572.27	584.43	591.92	599.24
	605.39	616.18	628.6	643.2	656.34	668.97	675.75	687.44	694.17	702.28	709.41	715.4
	726.91	734.62	748.81	761.23	790.85	804.14	816.73	831.08	844.45	857.46	871.31	885.86
FWHM (nm)	7.79	10.57	15.86	19.82	20.11	19.23	20.53	20.69	22.75	16.64	15.35	19.82
	26.55	26.72	30.81	28.61	27.9	28.98	27.85	30.01	30.59	28.29	25.45	26.13
	29.94	31.34	28	29.6	27.65	25.13	27.97	28.6	28.41	30.68	32.75	29.52

The UAV imagery was acquired during leaf-off (9 and 11 May) and leaf-on (29 June) 2016 in three separate flights in both seasons. The weather conditions were bright and cloudless during leaf-off campaigns and varied from sunny to cloudy during leaf-on campaigns (Table 3). The flight height was 100 m from the ground level, which provided a ground sampling distance (GSD) of 10 cm for the FPI and 2.5 cm for the RGB images. The flight speed was 3 m/s. The forward and side overlaps were 83% and 80%, respectively, for the FPI camera blocks and 96% and 85%, respectively, for the RGB camera blocks. Altogether, 20 ground control points (GCPs) were installed in the areas for georeferencing purposes (6 GCPs in YoS and 7 GCPs in both AdS east and west). They were targeted with circular targets with a 30-cm diameter, and their coordinates were measured using a Trimble R10 (L1 + L2) RTK-GPS receiver with accuracies of 2 cm in horizontal coordinates and 3 cm in height. For the reflectance transformation purposes, reflectance panels with a size of 1 m × 1 m and nominal reflectance of 0.03, 0.10, and 0.50 [34] were positioned near the UAV take-off place. An ASD Field Spec Pro (Analytical Spectral Devices, Malvern Panalytical Ltd., Malvern, United Kingdom) with cosine collector optics was installed near the take-off place to make irradiance measurements during the flights.

Table 3. Details of the UAV data capture in young seedling (YoS) and advanced seedling (AdS): date, time, sun zenith (SunZen) and azimuth (SunAz) angles, illumination conditions, and information about radiometric model used for FPI image processing (BRDF = bidirectional reflectance distribution function correction, RELA = relative image-wise corrections).

Spot	YoS		AdS West		AdS East	
Season	**Leaf-Off**	**Leaf-On**	**Leaf-Off**	**Leaf-On**	**Leaf-Off**	**Leaf-On**
Date	11 May	29 June	9 May	29 June	9 May	29 June
Time (UTC + 3)	11:41	15:11	12:10	13:57	11:31	13:12
SunZen	46°	42°	45°	38°	47°	38°
SunAz	148°	218°	158°	193°	145°	176°
Illumination Conditions	Bright	Bright	Bright	Variable	Bright	Overcast
Radiometric Model	BRDF	BRDF	BRDF	RELA	BRDF	RELA

2.3. Creating Dense Point Clouds and Image Mosaics

Georeferencing of the RGB images was carried out using the Pix4D MapperPro (Pix4D S.A., Prilly, Switzerland) version 2.2.25 software and supported by GCP and GNSS trajectory data collected on-board the UAV. After orientation processing, dense three-dimensional (3D) point clouds were created by automatic image matching using average point densities of 1600 points/m^2. Orientations of the FPI images were determined in a separate process. First, the orientations of three reference bands (band 3: L0 = 513.5 nm; band 11: L0 = 591.9; band 14: L0 = 616.2 nm) were calculated using the Pix4D software, as was the case with the RGB images. The rest of the bands were co-registered to the reference bands using a rigorous 3D approach [35]. The process provided the band registration with a better than 1-pixel accuracy over the area.

The objective of the radiometric processing of the FPI imagery was to provide high-quality reflectance mosaics including the 36 spectral bands. The radiometric modelling approach developed at the FGI and implemented in the FGI's radBA software (version 2016-08-20, Masala, Finland), an in-house toolbox for radiometric block adjustment [32,36], included sensor correction, atmospheric correction, correction for the illumination changes and other non-uniformities, and normalisation of the anisotropy effects due to the varying illumination and viewing directions [32]. The empirical line method [37] was used to calculate the transformation from digital numbers to reflectance factors with the aid of the reflectance reference panels. A radiometric block-adjustment method was used to determine the model-based radiometric correction to compensate for the radiometric disturbances. In this investigation, the relative image-wise correction parameters were calculated for all six datasets. Furthermore, disturbances caused by the object-reflectance anisotropy (i.e., bidirectional reflectance distribution function (BRDF)) were determined for the datasets that were collected during sunny weather (Table 3). Markelin et al. [38] previously evaluated different options of the radiometric processing.

The RGB image orthomosaics were calculated with a GSD of 2.5 cm using the Pix4D software. The hyperspectral orthomosaics were created using the FGI's radBA software ([32,36] with a GSD of 10 cm).

2.4. Delineation of Tree Crowns and Extracting 3D Metrics

The height of the photogrammetric point clouds was normalised to the height above-ground level using the national digital terrain model (DTM) with the resolution of 2 m. The DTM was created by the National Land Survey of Finland using ALS data. The expected elevation accuracy of the DTM varies from 10 to 30 cm in boreal forest conditions [39]. The DTM has been updated in August 2015.

For detecting tree crowns, leaf-off and leaf-on canopy height models (CHMs) were created from the normalised point clouds, by assigning the height value of the highest point to pixels of the CHMs. The resolution of 10 cm was selected for the CHMs to also match with the resolution of the FPI hyperspectral images. To avoid any empty pixels (gaps) in the CHMs, values for the null pixels were interpolated using the K-nearest neighbour inverse distance weighting (KNNidw) with the three nearest neighbours in the lidR package [40] in R 3.3.3 [41]. To delineate tree crowns from the CHMs of each plot, we applied watershed segmentation in SAGA GIS version 2.3.2 [42].

The maximum and mean height (H_{max}, H_{mean}) of segments were extracted from CHMs. Then, segments with H_{max} below a height threshold (0.5 m and 1.0 m for YoS and AdS, respectively) were excluded as ground vegetation or understory. Næsset and Bjerknes [26], and Økseter et al. [27], excluded ALS points below the 0.5-m threshold, assuming them to be laser returns from the ground vegetation. Therefore, we selected the threshold 0.5 for YoS and 1m AdS. Moreover, according to field data (Table 1), the smallest tree had H_{min} of 0.73 m in YoS and 1.57 m in AdS. Therefore, the selected thresholds were lower to include all tree segments. Within segments, we kept cells with height of $\geq 50\%$ of segments H_{max} to minimise the possible effect of understory. The 50% was selected by expert knowledge and visual inspection, although we admit that there might be other approaches.

2.5. Selection of Training Segments

The exclusion of segments with H_{max} below height thresholds was also applied for segments located outside sample plots boundary. Then, training segments were selected by visual interpretation of the well-distinguishable and typical leaf-off and leaf-on segments, located within a 2 m buffer around the sample plots boundary. The visual interpretation was carried out using leaf-off and leaf-on RGB orthomosaics to detect tree classes (spruce and birch) and non-tree classes (stumps/deadwood, bush/grass and rock). The number of training segments were 144 in leaf-off and 279 in leaf-on (Table 4). The non-tree classes were merged for the classification step.

Table 4. Number of training data in each classification class in each epoch. Non-tree classes include stump/deadwood, bush/grass/fern, and rock.

	Birch	Spruce	Non-Trees	Total
Leaf-off	30	67	47	144
Leaf-on	50	101	128	279

The number of training data is higher in leaf-on (Table 4) because grass, bushes, and ferns (which are in the non-tree class) had more segments in leaf-on data. They emerge in summer, grow fast, and reach the height thresholds.

2.6. Vegetation Indices and Finding Optimal Bands

We calculated the arithmetic mean of spectral values for each band from the hyperspectral data for each segment in the leaf-off and leaf-on data separately. These were then used to calculate three vegetation indices (VIs) (Equation (1)) using a combination of near-infra red (NIR) together with green (i), red (ii), and red-edge bands (iii). As there were several bands within these ranges of spectrum in our hyperspectral data (Table 5), we calculated all possible combinations of the three VIs using Equation (1).

$$\text{Index} = \frac{(R_{\lambda 1} - R_{\lambda 2})}{(R_{\lambda 1} + R_{\lambda 2})}, \tag{1}$$

where R is the reflectance value and $\lambda 1$ and $\lambda 2$ are the wavelengths of the two bands employed in the index.

To select the optimal and most important bands for the VIs in identifying trees from non-trees as well as spruce from birch, we ran the random forest method (implemented from yaImpute package [43] in R 3.3.3) 100 times for both leaf-off and leaf-on data. The final selection for the VIs was done based on the scaled importance; in other words, the VIs that were considered the most important variable at least 20 times of the 100 random forest runs were selected for the final modelling.

Table 5. Wavelength range and corresponding number of bands from hyperspectral images.

	Wavelength Range (nm)	Number of Bands
Green	507–562	8
Red	620–700	7
Red Edge	700–780	7
NIR	780–886	8

2.7. Segments Classification

In addition to the optimal VIs, H_{max} and H_{mean} were also used as predictors in training and prediction phases of the random forest classification method [44] to predict the species class of segments. The random forest classification method was applied to find the nearest neighbours (i.e., crown segments) in a feature space using the predictors selected (i.e., VIs, H_{max}, and H_{mean}).

We used the random forest method from the *yaImpute* R package with the *buildClasses* mode with 500 trees, k = 1, and we set the classification classes (birch, spruce, non-tree classes) as the *y* variable.

After classifying the segments, we discarded non-tree classes and proceeded to extract the plot-level total and spruce-specific TPH, as well as the mean height of all trees and of spruce trees' mean height. Note that tree heights were derived from CHMs and not predicted with the random forest method.

2.8. Accuracy Evaluation for Tree Density and Height

We compared plot-level spruce-specific TPH and the total TPH attributes with field-measured reference. To evaluate the reliability of remotely sensed tree height, we compared our estimation of the plot-level mean tree height with its corresponding field data, either spruce-specific tree heights or total tree heights, using the equations. Absolute and relative bias (BIAS) and RMSE were calculated for each attribute (Equations (2)–(5)).

$$\text{BIAS} = \frac{\sum_{i=1}^{n}\left(y_i - \hat{y}_i\right)}{n}, \tag{2}$$

$$\text{BIAS\%} = 100 \times \frac{\text{BIAS}}{\overline{y}}, \tag{3}$$

$$\text{RMSE} = \sqrt{\frac{\sum_{i=1}^{n}\left(y_i - \hat{y}_i\right)}{n}}, \tag{4}$$

$$\text{RMSE\%} = 100 \times \frac{\text{RMSE}}{\overline{y}}, \tag{5}$$

where n is the number of plots, y_i the value from the field data for plot i, $\hat{y}_i$ the remotely sensed (predicted) value for plot i, and $\overline{y}$ is the mean of the variable in the field data.

In addition, we also used Pearson correlation coefficient (r). The output value can be interpreted as the proportion of the variance in an attribute (remotely-sensed data) to the variance in another (field data) as x and y, respectively. The formula is the following:

$$r = \frac{\sum_{i=1}^{n}(x_i - \overline{x}) \cdot \left(y_i - \overline{y}\right)}{\sqrt{\sum_{i=1}^{n}(x_i - \overline{x})^2 \cdot \sum_{i=1}^{n}\left(y_i - \overline{y}\right)^2}}, \tag{6}$$

The accuracy evaluation for TPH and tree height was analysed and reported for all sample plots (n = 15) for both leaf-off and leaf-on conditions. Additionally, the accuracy was assessed among YoS (n = 5) and AdS (n = 10) separately.

3. Results

3.1. Analysing Spectral Features and Optimal Bands for Vegetation Indices

The spectral reflectance of training data using leaf-off and leaf-on hyperspectral data (Figure 2) showed that the tree classes are distinguishable from the non-tree class, especially in the red-edge and NIR spectrum in both epochs. The reflectance spectra from AdS leaf-on datasets had some anomalies (Figure 2b). The datasets were captured under cloudy or partially cloudy conditions using a short exposure time of 10 ms, which resulted in poor image quality, especially in the red spectral range (600–670 nm). This was not considered a critical problem in the analysis because only one of the indices was in this range, and our selection procedure did not select the lowest quality bands for the indices.

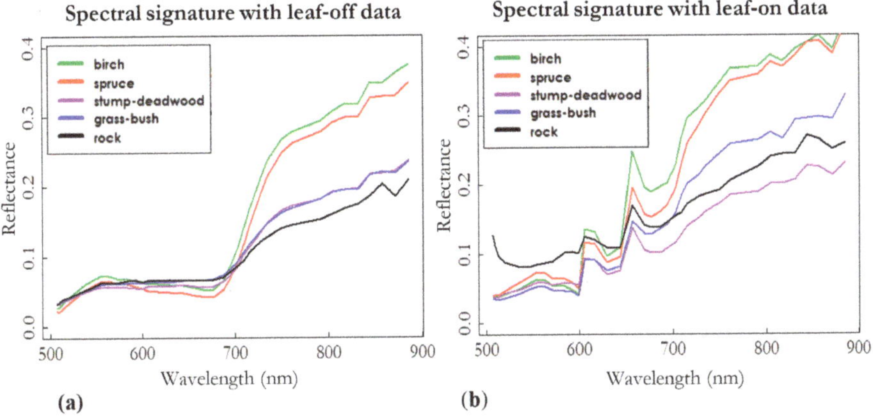

Figure 2. Mean spectra of training data in five classes, in leaf-off (**a**) and leaf-on (**b**).

The optimal bands found and used for creating VIs are given in Table 6 for each epoch.

Table 6. Wavelengths used for calculating vegetation indices.

Vegetation Index	Leaf-Off Wavelengths (nm)	Leaf-On Wavelengths (nm)
Red and NIR	694.16 and 857.46 675.75 and 804.15	675.75 and 871.31
Green and NIR	520.44 and 857.46 513.48 and 871.31	537.16 and 790.85
Red Edge and NIR	709.41 and 790.85 702.28 and 844.45 761.23 and 831.08	709.41 and 885.86 715.40 and 871.31 715.40 and 885.86 748.81 and 844.45

3.2. Tree Density Estimation

Both total tree density and spruce tree density were more accurately predicted with leaf-on data (Table 7). The total tree density was estimated approximately 10%-points more accurately (relative RMSE of 33.5% and 26.8% in leaf-off and leaf-on) than spruce tree density (44.6% and 38.1%) in both epochs.

The estimation of the total TPH was also compared among YoS and AdS separately (Figure 3). Total tree density of YoS were estimated more accurately with leaf-on data (relative RMSE of 32.7%) than with leaf-off data (relative RMSE of 47.3%). The total TPH was underestimated by 15.4% in leaf-on conditions whereas the underestimate for leaf-off conditions was 6.3%; although there was no substantial difference in relative and absolute RMSE between the epochs (Figure 3).

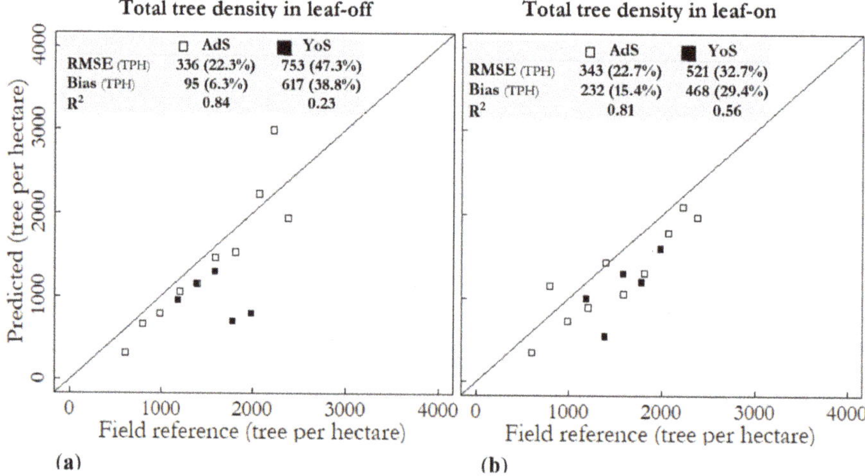

Figure 3. Total tree density (unit: TPH) in leaf-off (**a**) and leaf-on (**b**) conditions, separating plots in advanced seedling stand (AdS) and plots in young seedling stand (YoS).

Moreover, spruce tree density among YoS ($n = 5$) and AdS ($n = 10$) was also calculated for both epochs (Figure 4). The relative RMSE of spruce tree density in AdS was 19.2% in leaf-on data whereas it was 58.5% and 58.2% in YoS for both epochs. Spruce tree density was less underestimated in AdS in leaf-on (28.3% and 12.7% in leaf-off and leaf-on); nevertheless, it was approximately 4%-points more accurate in leaf-off in YoS (53.8% and 57.5% in leaf-off and leaf-on conditions).

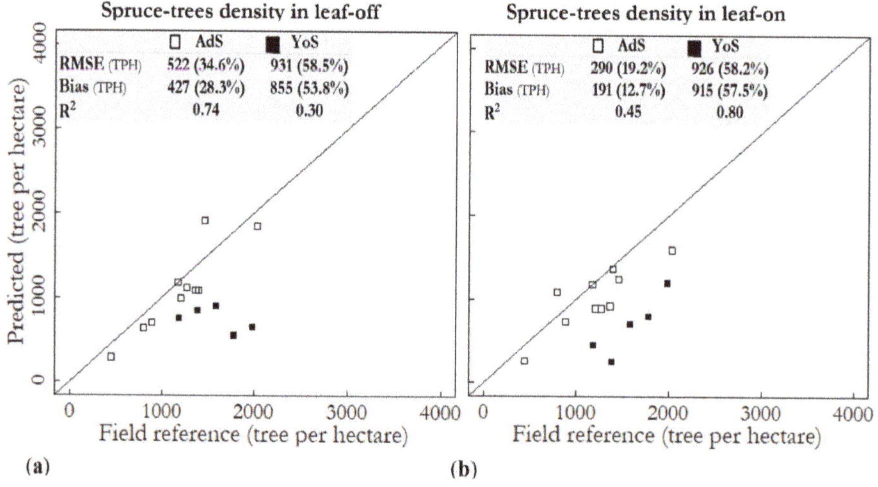

Figure 4. Spruce tree density (TPH) in leaf-off (**a**) and leaf-on (**b**) conditions, separating plots in advanced seedling stand (AdS) and plots in young seedling stand (YoS) in leaf-off and leaf-on conditions.

Table 7. Evaluation of the total and spruce tree densities among all sample plots ($n = 15$) (TPH = Trees per hectare).

	Total Number of Trees		Number of Spruce Trees	
	Leaf-Off	Leaf-On	Leaf-Off	Leaf-On
RMSE (TPH)	514	411	686	585
Relative RMSE (%)	33.5	26.8	44.6	38.1
Bias (TPH)	269	311	570	432
Bias %	17.5	20.2	37.1	28.1
R^2	0.57	0.73	0.46	0.35

3.3. Height Attribute Extraction

Among all sample plots, the mean height of all trees was underestimated by 20.8% and 7.4% (relative RMSE of 23.0% and 11.5%) in leaf-off and leaf-on conditions, respectively (Table 8). The mean height of spruces was underestimated by 20.2% and 6.9% (relative RMSE of 21.7% and 11.4%) with leaf-off and leaf-on data, respectively. As the results show, leaf-on data were more favourable for both all trees and the spruce mean height estimation. The absolute RMSEs and biases were 0.29 m and 0.18 m, respectively, for the leaf-on data, and 0.57 m and 0.52 m, respectively, for the leaf-off dataset.

Table 8. Evaluation of the total and spruce mean tree heights among all sample plots ($n = 15$).

	Mean Height of all of the Trees		Mean Height of Spruce Trees	
	Leaf-Off	Leaf-On	Leaf-Off	Leaf-On
RMSE (m)	0.57	0.29	0.52	0.27
Relative RMSE (%)	23.0	11.5	21.7	11.4
Bias (m)	0.52	0.18	0.48	0.16
Bias%	20.8	7.4	20.2	6.9
R^2	0.95	0.96	0.97	0.95

Figure 5 shows the estimation of the total tree height among YoS and AdS separately in both epochs. Leaf-on data resulted in more accurate estimations in both YoS and AdS. The mean height of all trees in AdS was underestimated by 20.1% in leaf-off, whereas it was improved to 8.3% in the leaf-on data. The underestimation in YoS was improved from 24.6% in leaf-off conditions to 2.6% in leaf-on conditions.

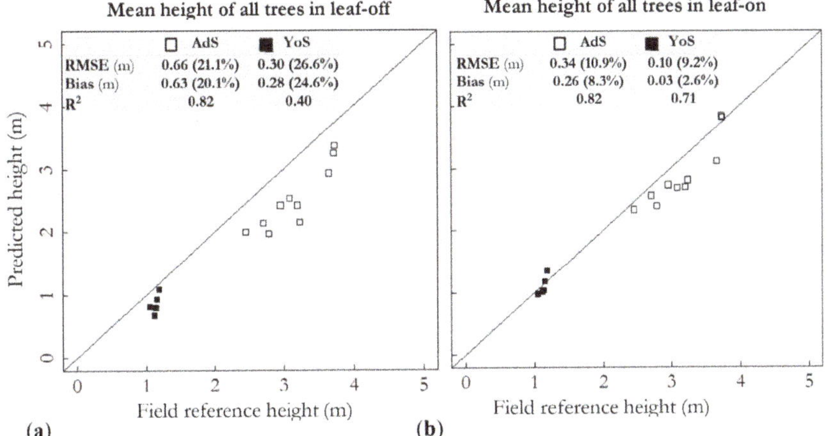

Figure 5. Mean tree height (unit: meter) of all trees in leaf-off (**a**) and leaf-on (**b**) conditions, separating plots in advanced seedling stand (AdS) and plots in young seedling stand (YoS).

The mean height of spruces was underestimated by 19.4% and 8.7% (relative RMSE of 19.8% and 10.8%) in the AdS in leaf-off and leaf-on conditions, respectively (Figure 6). Although it was underestimated by 24.6% (relative RMSE of 26.4%) for the YoS in leaf-off conditions; it was overestimated by 2.4% in leaf-on conditions (relative RMSE of 9.6%). The overestimation (Figure 6b) could be due to higher underestimation of spruce tree density in leaf-on (Figure 4b), which could show the omission of small spruce trees and result in the 2.4% overestimation. Relative RMSE for AdS (10.8%) was larger than YoS (9.6%) in leaf-on, in contrast to leaf-off conditions.

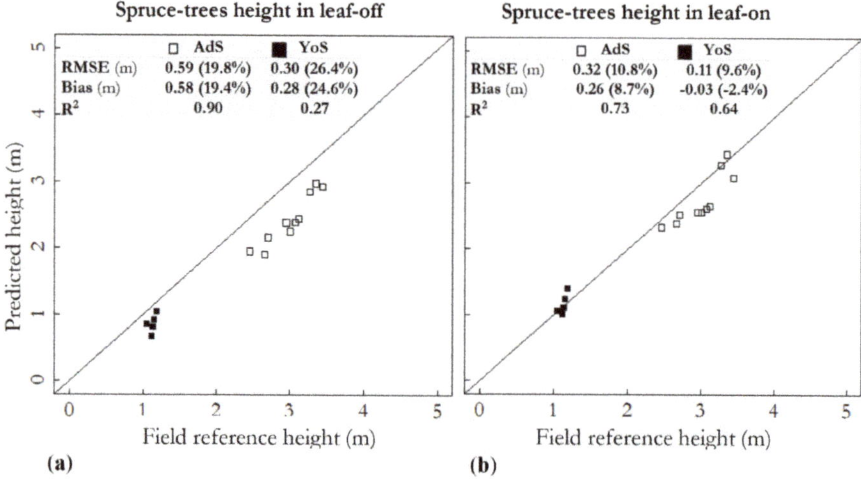

Figure 6. Spruce-specific mean tree height (unit: meter) in leaf-off (**a**) and leaf-on (**b**) conditions, separating plots in advanced seedling stand (AdS) and plots in young seedling stand (YoS) in leaf-off and leaf-on conditions.

4. Discussion

4.1. Tree Density Estimation

Our findings for total trees density in leaf-on (relative RMSE: 26.8%) was an improvement to [13] that achieved plot-level relative RMSE of 36.3%, that used area-based approach to fit random forest models with plot data and UAV for estimating forest attributes. We shall note that Puliti et al. [13] presented RMSE of different tree densities ranging between 1 to >10,000 at every 1000 intervals. Comparing the range of our field tree density (600–2400 TPH) with the corresponding reported interval in their results, our total tree leaf-on RMSE was more accurate (411 TPH) than their achievement (~1900 TPH). Our underestimation of tree density (leaf-off 17.5% and 20.2% leaf-on) was greater than [14] (13.6%). The greater underestimation can be because of different tree detection method they used three-step object-based methods, unlike our watershed-segmentation.

Comparing our findings with seedling-focused ALS studies, our relative RMSE was more accurate (26.8%) than [13] (53.4%). They used ALS data with point density of 5 points m^{-2} with the same methodology that they used for UAV data, area-based approach and random forest model fitting. Our higher point density and the different used method could consequence the outperformance. Earlier, Ørka et al. [25] applied ALS for predicting the attributes of 19 regeneration stands achieved a relative RMSE of 47% for predicting the total TPH at the stand level. Comparing our plot-level results with that of the stand level predictions in [25], our findings are more accurate because a decrease in the RMSE values was observed when scaling the plot-level estimation to stand level [13]. Moreover, an earlier study [26] utilised small-footprint ALS to estimate the tree height and the TPH in young forest stands (tree heights < 6 m). They resulted in a relative RMSE of 42% for predicting the stem number using a regression model, created with a combination of field reference data. Our tree density results (relative RMSE: 26.8%) were more reliable than above-mentioned studies in the seedling stands. However, we admit that every study can have various parameters that affect the results, such as sample size, tree species, density and height conditions, different resolution and quality of remote sensing imagery. The most comparable studies are [13,14], because the other above-mentioned literature used models to predict TPH, instead of direct detection of the trees from the remote sensing data.

4.2. Tree Height Estimation

Our findings for total trees leaf-on height (relative RMSE: 26.8% TPH) was in line with [13], achieved plot-level relative RMSE of 30.9 % using area-based approach and fitting random forest models with plot data and UAV to predict forest attributes. Our tree height estimation was more accurate than [17] that used UAV-based photogrammetric point clouds to assess the effects of the European spruce bark beetle (*Ips typographus* L.) disturbance on natural regeneration and standing deadwood. They reached a mean RMSE of 1.31 m (59%) and 1.57 m (64%) for manually and automatically delineated regeneration trees, respectively. They reported more accurate tree height estimation with point clouds from UAV than from aerial photography (mean relative RMSE of 115% and 59%, respectively), when manually delineating trees in both data. Moreover, Vepakomma et al. [16] resulted in an underestimation of 0.39 m in seedling tree height retrieval using UAV-based photogrammetric point clouds. Furthermore, Goodbody et al. [15] assessed the conditions of regeneration stands using digital aerial photography and UAVs, and underestimated tree height by 0.55 m (RMSE = 0.92 m) using UAV-based photogrammetric point clouds. They claimed that their result had the potential to be used in silvicultural prescriptions and growth projection models.

ALS has been another important data source for estimating tree height. Puliti et al. [13] achieved 32.0% of relative RMSE when assessing seedling tree height using 5 points m^{-2} density ALS data. Also, Ørka et al. [25] utilised ALS data in 19 regenerations stands in Norway for predicting tree attributes. They revealed relative RMSE of 28% for predicting the mean tree height at the stand level. Næsset and Bjerknes [26] predicted the plot-level height with 0.23 m (3.5%) of bias using a two-stage procedure. Note that only 29 sample plots (of the total 174 sample plots of their whole study area) were young stands (heights < 11.5m). Also, the tree height in the study [26] was higher than this study, although their tree density was higher (mean density 4197 TPH). Their smaller bias could be due to the two-stage procedure or because they had a mixture of mature stands in their study.

In our evaluations, the absolute RMSE and bias for mean tree height of all trees were 0.29 m and 0.18 m, respectively, for the leaf-on data, and 0.57 m and 0.52 m, respectively, for the leaf-off dataset. The values of the leaf-on data were close to the limits of the methods when considering the georeferencing accuracy of approximately 0.05 m, reconstruction accuracy of the tree surfaces of decimetres, and the uncertainty of the ALS based DTM, of approximately 0.10–0.30 m. The poorer accuracy of the leaf-off data is likely to be due to the challenges of 3D object reconstruction of leafless branches with of data set with a GSD of 2.5 cm using image matching; furthermore, the overall accuracy of the photogrammetric processing could be poorer with the more challenging leaf-off dataset. These results were better than in earlier studies for seedling stands although the earlier studies can have different parameters that can influence on the result such as sample size, tree species, density and height conditions, in addition to the quality of remotely-sensed data. The mean tree height in [17] varied between 1.19 and 4.10 m within eight sample plots, scanned by UAV at 40-m flying altitude. The GSD after optimisation process in the study had yielded average residuals < 10 cm in all plots (they used only RGB, not hyperspectral). Yet the flight height in this study was 100 m, which resulted in a GSD of hyperspectral data up to 10 cm and RGB-orthomosaics of 2.5 cm. The regeneration density in the study [17] varied more (approximately 300–8000 TPH) than in this study (approximately 1200–2000 TPH in YoS and 600–2400 TPH in AdS). Yet in this study, the deviation of plot-level tree height was higher (0.77–4.54 m) than in [17] that varied between 1.19–4.10 m.

In this study, the tree height could be even more accurately estimated if our field reference data were measured at the individual tree level or if at least the training data had tree-wise field-measured height. It is common that the tree height is also predicted at the same time as species classification carries out using the random forest method because it can handle predicting several attributes. This could improve the height estimation further and correct the small overestimation of tree height in YoS in the leaf-on data. Additionally, underestimation in tree density can cause overestimation in height retrieval, especially the omission of smaller trees. It was perhaps why the underestimation of spruce density in leaf-on caused a minor height overestimation (2.4%).

4.3. Comparing Leaf-Off and Leaf-On Data

This research was specifically designed to evaluate the performance of data collected in leaf-off and leaf-on conditions for seedling stands. It was observed that inventorying in leaf-on conditions is more favourable for both tree density and mean tree height overall, and we recommend the use of leaf-on data when object reconstruction is based on photogrammetry. Mean tree height was predicted more reliably (relative RMSE: 11.5%) than tree density (relative RMSE: 26.8%) among all sample plots with leaf-on data.

To the best of our knowledge, there were no prior literature to compare leaf-off and leaf-on data for characterizing seedling stands (using any type of remote sensing data, either from UAV, aerial imagery, active sensing or spaceborne). Therefore, we should compare our findings with studies that focused on non-seedling stands. In mature forests, leaf-off and leaf-on aerial images had been used to assess mapping of forest attributes [45]. They recommended against of using leaf-off aerial images, where coniferous trees (pine and spruce) were major species with birch trees as minor tree, because they observed poor accuracy and underestimation of height distribution using leaf-off data in deciduous forest. The lower height value estimation in leaf-off data was also reported by [46], that used leaf-off and leaf-on aerial images to estimate the proportion of deciduous stem volume in mixed coniferous-deciduous forest using area-based approach. Our findings are in parallel with their results. It is worth noting that further advantage of the leaf-on data includes the prospects of utilizing the spectral information in characterizing the vegetation.

In terms of ALS data for mature forests, small difference (relative RMSE and bias < 2%) was reported for estimating all forest attributes except volume (< 7%) between leaf-off and leaf-on data, affirming that leaf-off ALS data could be used for area-based methods [47]. Similarly, Lorey's mean height were estimated more accurately in leaf-off (RMSE: 0.07 m), than leaf-on (RMSE: 0.09 m) using area-based approach with ALS in a mixed managed boreal forest [48]. Also, other ALS studies recommended the use of leaf-off [49,50]. The reported slight advantage of leaf-off data in the ALS studies could be due to the used single-spectral ALS sensor that can be insufficient for discriminating different species in leaf-on conditions; in contrast to multi- and hyper-spectral data that outperformed in leaf-on in our research as well as other studies [45,46]. We note that further studies are required to examine this with more sample plots.

Our study showed that UAV imagery can be reliable used for characterizing seedling stands and may be a supplement or replacement for inventorying seedling stands in the future. Admittedly, further studies with more sample plots containing more deviation in tree height and density are required.

5. Conclusions

We used UAV-based photogrammetric point clouds and hyperspectral data to characterize tree attributes in seedling stands with our predesigned tree density and the species proportions of each field plot. Our data were acquired with leaf-on and leaf-off conditions.

Tree density feature in AdS were more accurately predicted compared to YoS, although tree density was higher in AdS. Thus, it can be concluded that the YoS (average height of less than 1.3 m) remained challenging to UAV-based photogrammetric point clouds and hyperspectral data and required further studies with more sample plots.

Overall, mean tree height of all and spruce trees were estimated more accurately in leaf-on conditions for both YoS and AdS. Comparing both height estimations between YoS and AdS in leaf-on conditions, the heights were estimated more accurately for YoS than AdS.

Comparing epochs, it can be concluded that collecting remotely sensed data in leaf-on conditions could be more favorable because our findings showed lower absolute and relative RMSE with leaf-on data for both the total and spruce tree density. The superiority of the leaf-on condition, considering absolute and relative RMSE, was the same for mean tree height of all and spruces trees, for both YoS and AdS, although some absolute and relative bias were different. Generally, leaf-on data is recommended especially when using photogrammetric reconstruction method, and furthermore, when using hyperspectral data, the leaf-on data might provide further information about the condition of the vegetation.

Author Contributions: Conceptualization, N.S., E.H., M.H., J.H. and M.V.; Data curation, M.I.; Formal analysis, M.I.; Funding acquisition, E.H., M.H., J.H. and M.V.; Methodology, M.I., M.H. and M.V.; Project administration, M.H. and J.H.; Resources, L.M., T.R., R.N., T.H. and E.H.; Supervision, N.S., E.H., M.H. and M.V.; Writing–original draft, M.I.; Writing–review & editing, M.I., N.S., L.M., T.R., R.N., T.H., E.H., M.H., J.H. and M.V.

Funding: The research was funded by the Academy of Finland through the project "Unmanned Airborne Vehicle-based 4D Remote Sensing for Mapping Rain Forest Biodiversity and Its Change in Brazil" (Project number 292941), Finnish Ministry of Agriculture and Forestry (project number 188/03.02.02.00/2016), and the Centre of Excellence in Laser Scanning Research (project number 307362).

Acknowledgments: The authors would like to thank Tomi Koivisto for establishing the field plots by thinning trees to target densities and collecting the field data. Additionally, Häme University of Applied Sciences and Evo field station are thanked for supporting our research activities at Evo study site.

Conflicts of Interest: The authors declare no conflict of interest. The funders had no role in the design of the study; in the collection, analyses, or interpretation of data; in the writing of the manuscript, or in the decision to publish the results.

References

1. Puliti, S.; Ene, L.T.; Gobakken, T.; Næsset, E. Use of partial-coverage UAV data in sampling for large scale forest inventories. *Remote Sens. Environ.* **2017**, *194*, 115–126. [CrossRef]

2. Goodbody, T.R.H.; Coops, N.C.; Marshall, P.L.; Tompalski, P.; Crawford, P. Unmanned aerial systems for precision forest inventory purposes: A review and case study. *For. Chron.* **2017**, *93*, 71–81. [CrossRef]

3. Torresan, C.; Berton, A.; Carotenuto, F.; Di Gennaro, S.F.; Gioli, B.; Matese, A.; Miglietta, F.; Vagnoli, C.; Zaldei, A.; Wallace, L. Forestry applications of UAVs in Europe: A review. *Int. J. Remote Sens.* **2016**, *38*, 1–21. [CrossRef]

4. Zahawi, R.A.; Dandois, J.P.; Holl, K.D.; Nadwodny, D.; Reid, J.L.; Ellis, E.C. Using lightweight unmanned aerial vehicles to monitor tropical forest recovery. *Biol. Conserv.* **2015**, *186*, 287–295. [CrossRef]

5. Tuominen, S.; Balazs, A.; Saari, H.; Pölönen, I.; Sarkeala, J.; Viitala, R. Unmanned aerial system imagery and photogrammetric canopy height data in area-based estimation of forest variables. *Silva. Fenn.* **2015**, *49*, 1–19. [CrossRef]

6. Puliti, S.; Ørka, H.O.; Gobakken, T.; Næsset, E. Inventory of small forest areas using an unmanned aerial system. *Remote Sens.* **2015**, *7*, 9632–9654. [CrossRef]

7. Lisein, J.; Pierrot-deseilligny, M.; Lejeune, P.; Management, N. A Photogrammetric Workflow for the Creation of a Forest Canopy Height Model from Small Unmanned Aerial System Imagery. *Forests* **2013**, 922–944. [CrossRef]

8. Wallace, L.O.; Lucieer, A.; Watson, C.S. Assessing the feasibility of UAV-based LiDAR for high resolution forest change detection. In *Proceedings of the 12th Congress of the International Society for Photogrammetry and Remote Sensing, Melbourne, Australia, 25 August–1 September 2012*; International Archives of the Photogrammetry, Remote Sensing and Spatial Information Science: Melbourne, Australia, 2012; Volume 39, pp. 499–504.

9. Carr, J.C.; Slyder, J.B. Individual tree segmentation from a leaf-off photogrammetric point cloud. *Int. J. Remote Sens.* **2018**, *39*, 5195–5210. [CrossRef]

10. Thiel, C.; Schmullius, C. Comparison of UAV photograph-based and airborne lidar-based point clouds over forest from a forestry application perspective. *Int. J. Remote Sens.* **2017**, *38*. [CrossRef]

11. Tapio. *Hyvän Metsänhoidon Suositukset. (Recommendations for Forest Management in Finland)*; Forest Development Centre Tapio. Metsäkustannus oy: Helsinki, Finland, 2006; p. 100. (In Finnish)

12. Huuskonen, S.; Hynynen, J. Timing and intensity of precommercial thinning and their effects on the first commercial thinning in Scots pine stands. *Silva. Fenn.* **2006**, *40*, 645–662. [CrossRef]

13. Puliti, S.; Solberg, S.; Granhus, A. Use of UAV Photogrammetric Data for Estimation of Biophysical Properties in Forest Stands Under Regeneration. *Remote Sens.* **2019**, *11*, 233. [CrossRef]

14. Feduck, C.; Mcdermid, G.J. Detection of Coniferous Seedlings in UAV Imagery. *Forests* **2018**, *9*, 432. [CrossRef]

15. Goodbody, T.R.H.; Coops, N.C.; Hermosilla, T.; Tompalski, P.; Crawford, P. Assessing the status of forest regeneration using digital aerial photogrammetry and unmanned aerial systems. *Int. J. Remote Sens.* **2018**, *39*, 5246–5264. [CrossRef]

16. Vepakomma, U.; Cormier, D.; Thiffault, N. Potential of UAV based convergent photogrammetry in monitoring regeneration standards. *Int. Arch. Photogramm. Remote Sens. Spat. Inf. Sci.* **2015**, *40*, 281–285. [CrossRef]

17. Röder, M.; Latifi, H.; Hill, S.; Wild, J.; Svoboda, M.; Brůna, J.; Macek, M.; Nováková, M.H.; Gülch, E.; Heurich, M. Application of optical unmanned aerial vehicle-based imagery for the inventory of natural regeneration and standing deadwood in post-disturbed spruce forests. *Int. J. Remote Sens.* **2018**, *39*, 5288–5309. [CrossRef]

18. Kirby, C.L. *A Camera and Interpretation System for Assessment of Forest Regeneration*; Northern Forest Research Center: Edmonton, AB, Canada, 1980.

19. Hall, R.J.; Aldred, A.H. Forest regeneration appraisal with large-scale aerial photographs. *For. Chron.* **1992**, *68*, 142–150. [CrossRef]

20. Pouliot, D.A.; King, D.J.; Pitt, D.G. Development and evaluation of an automated tree detection—delineation algorithm for monitoring regenerating coniferous forests. *Can. J. For. Res.* **2005**, *35*, 2332–2345. [CrossRef]

21. Pouliot, D.A.; King, D.J.; Pitt, D.G. Automated assessment of hardwood and shrub competition in regenerating forests using leaf-off airborne imagery. *Remote Sens. Environ.* **2006**, *102*, 223–236. [CrossRef]

22. Wunderle, A.L.; Franklin, S.E.; Guo, X.G. Regenerating boreal forest structure estimation using SPOT-5 pan-sharpened imagery. *Int. J. Remote Sens.* **2007**, *28*, 4351–4364. [CrossRef]

23. Hauglin, M.; Bollandsås, O.M.; Gobakken, T.; Næsset, E. Monitoring small pioneer trees in the forest-tundra ecotone: Using multi-temporal airborne laser scanning data to model height growth. *Environ. Monit. Assess.* **2018**, *190*. [CrossRef]

24. Thieme, N.; Martin Bollandsås, O.; Gobakken, T.; Næsset, E. Detection of small single trees in the forest-tundra ecotone using height values from airborne laser scanning. *Can. J. Remote Sens.* **2011**, *37*, 264–274. [CrossRef]

25. Ole Ørka, H.; Gobakken, T.; Næsset, E. Predicting Attributes of Regeneration Forests Using Airborne Laser Scanning. *Can. J. Remote Sens.* **2016**, *42*, 541–553. [CrossRef]

26. Næsset, E.; Bjerknes, K.O. Estimating tree heights and number of stems in young forest stands using airborne laser scanner data. *Remote Sens. Environ.* **2001**, *78*, 328–340. [CrossRef]

27. Økseter, R.; Bollandsås, O.M.; Gobakken, T.; Næsset, E. Modeling and predicting aboveground biomass change in young forest using multi-temporal airborne laser scanner data. *Scand. J. For. Res.* **2015**, *30*, 458–469. [CrossRef]

28. Korpela, I.; Tuomola, T.; Tokola, T.; Dahlin, B. Appraisal of seedling stand vegetation with airborne imagery and discrete-return LiDAR—an exploratory analysis. *Silva. Fenn.* **2008**, *42*, 753–772. [CrossRef]

29. Korhonen, L.; Pippuri, I.; Packalén, P.; Heikkinen, V.; Maltamo, M.; Heikkilä, J. Detection of the need for seedling stand tending using high-resolution remote sensing data. *Silva. Fenn.* **2013**, *47*, 1–20. [CrossRef]

30. Näslund, M. *Skogsförsöksanstaltens gallrings-försök i tallskog. Meddelanden från Statens Skogs-försöksanstalt 29*; EDDELANDEN FRÅN STATENS KOGSFöRSöKSANSTALT HAFTE 29: Stockholm, Sweden, 1936; p. 169. (In Swedish)

31. Saari, H.; Pölönen, I.; Salo, H.; Honkavaara, E.; Hakala, T.; Holmlund, C.; Mäkynen, J.; Mannila, R.; Antila, T.; Akujärvi, A. Miniaturized hyperspectral imager calibration and UAV flight campaigns. In *Proceedings of the SPIE 8889, Sensors, Systems, and Next-Generation Satellites XVII, 88891O, Dresden, Germany, 24 October 2013*; Meynart, R., Neeck, S.P., Shimoda, H., Eds.; International Society for Optics and Photonics SPIE: Dresden, Germany, 2013.

32. Honkavaara, E.; Saari, H.; Kaivosoja, J.; Pölönen, I.; Hakala, T.; Litkey, P.; Mäkynen, J.; Pesonen, L. Processing and assessment of spectrometric, stereoscopic imagery collected using a lightweight UAV spectral camera for precision agriculture. *Remote Sens.* **2013**, *5*, 5006–5039. [CrossRef]

33. Honkavaara, E.; Hakala, T.; Nevalainen, O.; Viljanen, N.; Rosnell, T.; Khoramshahi, E.; Näsi, R.; Oliveira, R.; Tommaselli, A. Geometric and reflectance signature characterization of complex canopies using hyperspectral stereoscopic images from uav and terrestrial platforms. *Int. Arch. Photogramm. Remote Sens. Spat. Inf. Sci.* **2016**, *41*, 77–82. [CrossRef]

34. Honkavaara, E.; Markelin, L.; Hakala, T.; Peltoniemi, J. The Metrology of Directional, Spectral Reflectance Factor Measurements Based on Area Format Imaging by UAVs. *Photogramm.-Fernerkund.-Geoinf.* **2014**, *2014*, 175–188. [CrossRef]

35. Honkavaara, E.; Rosnell, T.; Oliveira, R.; Tommaselli, A. Band registration of tuneable frame format hyperspectral UAV imagers in complex scenes. *ISPRS J. Photogramm. Remote Sens.* **2017**, *134*, 96–109. [CrossRef]

36. Honkavaara, E.; Khoramshahi, E. Radiometric correction of close-range spectral image blocks captured using an unmanned aerial vehicle with a radiometric block adjustment. *Remote Sens.* **2018**, *10*, 256. [CrossRef]

37. Smith, G.M.; Milton, E.J. The use of the empirical line method to calibrate remotely sensed data to reflectance. *Int. J. Remote Sens.* **1999**, *20*, 2653–2662. [CrossRef]

38. Markelin, L.; Honkavaara, E.; Näsi, R.; Viljanen, N.; Rosnell, T.; Hakala, T.; Vastaranta, M.; Koivisto, T.; Holopainen, M. Radiometric correction of multitemporal hyperspectral uas image mosaics of seedling stands. *Int. Arch. Photogramm. Remote Sens. Spat. Inf. Sci.* **2017**, *42*, 113–118. [CrossRef]

39. Hyyppä, J.; Hyyppä, H.; Leckie, D.; Gougeon, F.; Yu, X.; Maltamo, M. Review of methods of small-footprint airborne laser scanning for extracting forest inventory data in boreal forests. *Int. J. Remote Sens.* **2008**, *29*, 1339–1366. [CrossRef]

40. Roussel, J.-R.; Auty, D. lidR: Airborne LiDAR Data Manipulation and Visualization for Forestry Applications. R package version 1.2.1. 2017. Available online: https://CRAN.R-project.org/package=lidR (accessed on 30 November 2017).

41. R Core Team (2017). R: A language and environment for statistical computing. R Foundation for Statistical Computing, Vienna, Austria. Available online: https://www.R-project.org/ (accessed on 12 September 2017).

42. Conrad, O.; Bechtel, B.; Bock, M.; Dietrich, H.; Fischer, E.; Gerlitz, L.; Wehberg, J.; Wichmann, V.; Böhner, J. System for Automated Geoscientific Analyses (SAGA) v. 2.1.4. *Geosci. Model. Dev.* **2015**, *8*, 1991–2007. [CrossRef]

43. Crookston Nicholas, L.; Finley Andrew, O. YaImpute: An R Package for k-NN Imputation. J. Stat. Softw. 2008. Available online: https://CRAN.R-project.org/package=yaImpute (accessed on 1 March 2018).

44. Breiman, L. Random forests. *Mach. Learn.* **2001**, *45*, 5–32. [CrossRef]

45. Bohlin, J.; Bohlin, I.; Jonzén, J.; Nilsson, M. Mapping forest attributes using data from stereophotogrammetry of aerial images and field data from the national forest inventory. *Silva. Fenn.* **2017**, *51*, 1–18. [CrossRef]

46. Bohlin, J.; Wallerman, J.; Fransson, J.E.S. Deciduous forest mapping using change detection of multi-temporal canopy height models from aerial images acquired at leaf-on and leaf-off conditions height models from aerial images acquired at leaf-on and leaf-off conditions. *Scand. J. For. Res.* **2016**, *31*, 517–525. [CrossRef]

47. White, J.C.; Arnett, J.T.T.R.; Wulder, M.A.; Tompalski, P.; Coops, N.C. Evaluating the impact of leaf-on and leaf-off airborne laser scanning data on the estimation of forest inventory attributes with the area-based approach. *Can. J. For. Res.* **2015**, *45*, 1498–1513. [CrossRef]

48. Næsset, E. Assessing sensor effects and effects of leaf-off and leaf-on canopy conditions on biophysical stand properties derived from small-footprint airborne laser data. **2005**, *98*, 356–370. [CrossRef]

49. Villikka, M.; Packalén, P.; Maltamo, M. The suitability of leaf-off airborne laser scanning data in an area-based forest inventory of coniferous and deciduous trees. *Silva. Fenn.* **2012**, *46*, 99–110. [CrossRef]

50. Ørka, H.O.; Næsset, E.; Bollandsås, O.M. Effects of different sensors and leaf-on and leaf-off canopy conditions on echo distributions and individual tree properties derived from airborne laser scanning. *Remote Sens. Environ.* **2010**, *114*, 1445–1461. [CrossRef]

 *forests* 

Article

The Use of Low-Altitude UAV Imagery to Assess Western Juniper Density and Canopy Cover in Treated and Untreated Stands

Nicole Durfee [1,*], Carlos G. Ochoa [2] and Ricardo Mata-Gonzalez [3]

1   Water Resources Graduate Program—Ecohydrology Lab, Oregon State University, Corvallis, OR 97331, USA
2   Department of Animal and Rangeland Sciences—Ecohydrology Lab, Oregon State University, Corvallis, OR 97331, USA; Carlos.Ochoa@oregonstate.edu
3   Department of Animal and Rangeland Sciences, Oregon State University, Corvallis, OR 97331, USA; ricardo.matagonzalez@oregonstate.edu
*   Correspondence: durfeen@oregonstate.edu; Tel.: +1-541-737-0933

Received: 12 March 2019; Accepted: 25 March 2019; Published: 29 March 2019

Abstract: Monitoring vegetation characteristics and ground cover is crucial to determine appropriate management techniques in western juniper (*Juniperus occidentalis* Hook.) ecosystems. Remote-sensing techniques have been used to study vegetation cover; yet, few studies have applied these techniques using unmanned aerial vehicles (UAV), specifically in areas of juniper woodlands. We used ground-based data in conjunction with low-altitude UAV imagery to assess vegetation and ground cover characteristics in a paired watershed study located in central Oregon, USA. The study was comprised of a treated watershed (most juniper removed) and an untreated watershed. Research objectives were to: (1) evaluate the density and canopy cover of western juniper in a treated (juniper removed) and an untreated watershed; and, (2) assess the effectiveness of using low altitude UAV-based imagery to measure juniper-sapling population density and canopy cover. Ground-based measurements were used to assess vegetation features in each watershed and as a means to verify analysis from aerial imagery. Visual imagery (red, green, and blue wavelengths) and multispectral imagery (red, green, blue, near-infrared, and red-edge wavelengths) were captured using a quadcopter style UAV. Canopy cover in the untreated watershed was estimated using two different methods: vegetation indices and support vector machine classification. Supervised classification was used to assess juniper sapling density and vegetation cover in the treated watershed. Results showed that vegetation indices that incorporated near-infrared reflectance values estimated canopy cover within 0.7% to 4.1% of ground-based calculations. Canopy cover estimates at the untreated watershed using supervised classification were within 0.9% to 2.3% of ground-based results. Supervised classification applied to fall imagery using multispectral bands provided the best estimates of juniper sapling density compared to imagery taken in the summer or to using visual imagery. Study results suggest that low-altitude multispectral imagery obtained using small UAV can be effectively used to assess western juniper density and canopy cover.

Keywords: juniper woodlands; ecohydrology; remote sensing; unmanned aerial systems; central Oregon; rangelands

---

1. Introduction

The range and density of woody plant species such as western juniper (*Juniperus occidentalis* Hook.) have substantially increased in the western United States over the last 150 years. Estimates of juniper (*Juniperus* spp.) expansion across the Great Basin range from 125% to 625% [1] and western juniper alone can be found across 3.6 million ha in the intermountain west [2]. The expansion of western juniper in particular has arisen in two primary forms: through the encroachment of juniper into areas previously dominated by sagebrush (*Artemisia tridentata* Nutt.), and through increases in the density of juniper in areas where it was relatively sparse [3]. Historically, juniper was largely found in areas with lower fire risk [4]. Intensive grazing, reduced fire occurrence, and favorable wetter climate conditions have all been cited as reasons for the vast juniper expansion observed in the late 19th and early 20th century [3,5].

Juniper expansion is a concern in many rangeland areas as it may lead to reduced water availability for other types of vegetation. Increased juniper canopy cover has been associated with increased bare ground and decreased shrub, forb, and grass cover [6] and reductions in vegetation production and diversity [7]. Several studies [6,8–11] have addressed the impacts of juniper expansion on ecological and hydrological processes. These impacts include increased erosion and runoff [12–14] and decreased soil moisture [8] typically associated with shifts in vegetation cover [2,12,15], particularly with increased bare ground in intercanopy locations [16].

The use of ground-based techniques to assess vegetation characteristics is limited to the resources available and normally includes an aggregate of data collected at point specific locations. Remote sensing offers the ability to assess ecological features over larger temporal and spatial scales. Remote sensing has been used successfully to identify juniper expansion [17], assess shrub cover characteristics in encroached sagebrush steppe ecosystems [18], calculate canopy cover in juniper woodlands [19,20] and characterize ground cover following treatment [21]. Remote-sensing techniques using multispectral imagery (particularly near-infrared reflectance) has improved the ability to assess changes in vegetation cover [22,23].

The use of unmanned aerial vehicle (UAV)-based data collection in remote-sensing applications can be useful in western juniper research including species classification [24], soil erosion monitoring [25], and measurements of tree canopy [26,27]. Imagery captured using remote sensing can improve our ability to study juniper removal and recovery by reducing time requirements for data collection and providing greater flexibility in observation times compared to ground-based measurements alone.

Vegetation indices derived from aerial and satellite imagery can be particularly useful for vegetation identification and classification as they provide information about vegetation characteristics by analyzing specific band reflectance properties. The relationship between different spectral bands can be used to distinguish between areas of vegetation and bare soil or rock. Vegetation indices developed from remote sensing data have been used to determine gross primary production in pinyon-juniper woodlands [28].

There is a range of vegetation indices used to assess vegetation characteristics. Reflectance characteristics from multispectral imagery, particularly the near-infrared and red-edge spectral regions, have been successfully used to assess vegetation growth [29] and to identify different vegetation species [24,30]. A commonly used vegetation index, the normalized difference vegetation index (NDVI) [31], is calculated from the reflectance characteristics of the near-infrared and red bands, and indicates photosynthetic activity. NDVI has also been found to be closely related to ground-based canopy cover calculations [32]. The optimized soil adjusted vegetation index (OSAVI) has a similar formula to that of NDVI but is used to minimize the influence of soil reflectance [33], a concern in many arid and semiarid regions with high amounts of bare soil. Another example of an index potentially useful in studying western juniper is the total ratio vegetation index (TRVI) [34], which was developed to address different vegetation characteristics in arid and semiarid ecosystems (e.g., juniper woodlands).

Other techniques used in image analysis such as classification have been utilized extensively for the detection and assessment of vegetation [35–37]. Pixel-based image analysis has been used for weed detection [38] and vegetation identification in mixed plant communities [39]. The use of classification tools for image analysis has also been shown to decrease time requirements compared to manual analysis of imagery [40], particularly over large areas. In ecosystems where juniper expansion is occurring, visual inspection of imagery alone is time-consuming and inefficient due to the typically large areas involved. By incorporating classification into analysis of UAV-based imagery, it may be possible to more efficiently monitor juniper re-establishment after removal.

This study sought to build upon UAV-based techniques and ground- based data collection to study the spatial distribution of juniper in a treated watershed where mature western juniper was eliminated in 2005 and in an untreated watershed. Study objectives were to: (1) evaluate the density and canopy cover of western juniper in a treated (juniper removed) and an untreated watershed; and, (2) assess the effectiveness of using low altitude UAV-based imagery to measure juniper-sapling population density and canopy cover.

2. Materials and Methods

2.1. Study Site

The study was conducted at the Camp Creek Paired Watershed Study (CCPWS) site in central Oregon. The CCPWS was established in 1993 to study long-term ecohydrological relationships in western juniper-dominated landscapes [41]. The study site includes two watersheds (WS): an untreated WS (96 hectares) and a treated WS (116 hectares) (Figure 1). During 2005 to 2006, nearly 90% of western juniper trees were removed from the treated WS. Trees were cut using chainsaws, the boles were removed, and tree limbs were scattered [42].

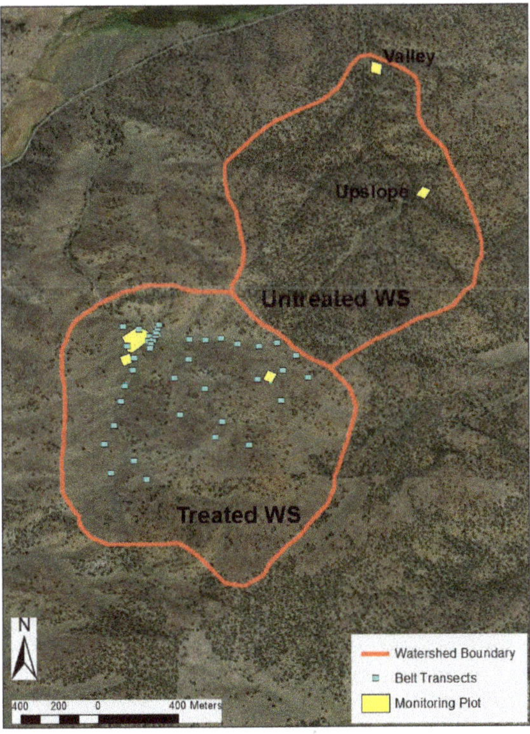

Figure 1. Map of the study site showing ground-based collection points and monitoring plot location in both watersheds. Study plots in the untreated watershed (WS) have been labeled to clarify position. The larger plot in the treated WS indicates the location of the unmanned aerial vehicle (UAV) imagery collection in that WS. Image created using ArcMap 10.6. Source: Esri, DigitalGlobe.

Climate in central Oregon is semiarid and precipitation falls largely during the fall and winter months. Average annual precipitation (2009–2017) at the study site is 358 mm [41]. Elevation at the study site ranges from 1350 to 1500 m. The orientation of the treated WS is primarily north by northwest while the untreated WS is largely oriented toward the north [43]. The average slope is 25% in the untreated WS and 24% in the treated WS [43].

Overstory vegetation at the treated WS is dominated by big sagebrush, while western juniper is the dominant species at the untreated WS. Understory vegetation in both watersheds is dominated by perennial grasses, primarily bluebunch wheatgrass [*Pseudoroegneria spicata* (Pursh.) Á. Löve], Idaho fescue (*Festuca idahoensis* Elmer), Indian ricegrass [*Achnatherum hymenoides* (Roem. and Schult.) Barkworth], Sandberg bluegrass (*Poa secunda* J. Presl), and Thurber's needlegrass [*Achnatherum thurberianum* (Piper) Barkworth] and some annual grasses, such as cheatgrass (*Bromus tectorum* L.) [43,44]. As reported by Ray et al. [44], juniper canopy cover at the untreated WS is 31% and at the treated WS is less than 1%, this based on surveys in 2015. According to the juniper occupancy classification described by Miller et al. [2], the untreated WS is considered at the highest level, Phase III, in which juniper is at nearly 30% occupancy and it is the dominant overstory species.

2.2. Vegetation Measurements

We calculated juniper-sapling population density, canopy cover, and age characteristics at the watershed scale in the treated WS, and adult and sapling-stage tree density and canopy cover at the plot scale in both watersheds.

At the watershed scale, we installed 41 belt transects (30 m by 3 m) to measure juniper-sapling count, height, and crown width in the treated WS. The belt transects were located across the landscape to represent varying aspect and slope characteristics. A subsample of 18 saplings representing common tree characteristics (i.e., height and width) observed in the treated WS were removed to determine tree age using techniques described by Phipps [45].

At the plot scale, we installed two 2000 m^2 monitoring plots in each watershed. One plot was installed in a valley location near the watershed outlet and one in an upstream location (Figure 1). In the untreated WS, tree canopy cover was estimated using a spherical concave densiometer (model A) (Forestry Suppliers, Jackson, MS, USA) across five 40 m parallel transects in each plot. Tree canopy cover was measured every 5 m, facing each cardinal direction, in each 40 m transect. The two plots in the untreated WS were also used to assess juniper canopy cover estimates using UAV-based imagery. All juniper (sapling and adult stages) were counted on the ground to determine tree population density in the monitoring plots at both watersheds. No adult-stage trees were present in either of the two plots in the treated WS; therefore, we only measured sapling count in each 2000 m^2 plot.

An 11,600 m^2 plot in the treated WS (Figure 1) was employed to assess juniper sapling estimates obtained using the various UAV-based methods described below. A subset of juniper saplings in the larger plot were selected to assess the accuracy of juniper identification.

2.3. Unmanned Aerial Vehicle (UAV)-Based Imagery Collection and Analysis

We collected UAV-based imagery from multiple low elevation (40 to 50 m) flights conducted on 21 June 2018, 15 and 16 July 2018, and 12 October 2018. In order to minimize shadows, flights occurred around noon and early afternoon. Aside from temperature, weather conditions at the time of each flight were similar with scattered clouds present and light, variable winds. Data collected in October 2018 and in the summer of 2018 were used to compare multispectral imagery results across seasons in the treated WS. Three quadcopter UAV (Table 1) were used to collect data. Multispectral imagery (red, green, blue, near-infrared, and red-edge) was captured using a RedEdge camera (MicaSense, Inc., Seattle, WA, USA). The RedEdge camera was attached either to a Matrice 100 (DJI, Shenzhen, China) or to a Solo (3D Robotics, Inc., Berkeley, CA, USA) UAV for multispectral imagery collection. RGB (red, green, and blue wavelengths) imagery was collected using a DJI Phantom 3 Professional camera (DJI, Shenzhen, China). The Phantom 3 was not used for image collection at the treated WS, the red, green, and blue bands were extracted from the multispectral raster and used for analysis of visual imagery instead.

Table 1. UAVs used for data collection. Multispectral (red, green, blue, near-infrared, and red-edge bands) imagery was captured by attaching the RedEdge camera to the Solo and Matrice 100. Visual imagery (red, green, and blue bands) was collected using the Phantom 3 Professional camera.

UAV Platform	Manufacturer	Image Type
Solo	3D Robotics, Inc., Berkeley, CA, USA	Multispectral
Matrice 100	DJI, Shenzhen, China	Multispectral
Phantom 3 Professional	DJI, Shenzhen, China	Visual

These quadcopter UAV were selected for use as they provide a flexible platform that can be adapted for multiple types of data collection. Additionally, given the remote location and lack of suitable launching and landing areas, a fixed wing aircraft would be difficult to use. The quadcopters used in this study offer the advantage of being relatively easy to operate, making them an ideal candidate to be used by land managers who may not have flight experience or access to more expensive UAV.

Flight plans were created and conducted using the Pix4Dcapture (Pix4D SA, Lausanne, Switzerland) mobile application, or flown manually. Relatively low flight altitudes (40 to 50 m above ground level) were employed in order to assess the ability of the UAV-imagery to detect juniper saplings. These flight altitudes were chosen as they provide relatively high spatial resolution, while requiring less flight time than much lower flight altitudes. While time requirements for UAV-based data collection were not intensive (flights averaged less than 12 min), the intent of this study was to assess a method that could be applied to larger study areas in the future. Additionally, the height of mature juniper in the untreated WS required a minimum of 40 m flight altitude to ensure sufficient clearance from the top of the trees. An overlapping grid pattern was flown to ensure minimum 80% forward overlap and 60% side overlap. Sufficient overlap was confirmed using the PhotoScan professional (Agisoft LLC, St Petersburg, Russia) program after individual images were added and aligned. All images were captured at nadir.

Orthomosaics created using RGB imagery had slightly higher spatial resolution than those created using the multispectral imagery. Spatial resolution of the visual band orthomosaic at the untreated valley site was 2.09 cm pixel^{-1} and 2.02 cm pixel^{-1} at the untreated upstream site. Spatial resolution of the multispectral imagery at the untreated WS valley site was 3.15 cm pixel^{-1} and 2.74 cm pixel^{-1} at the untreated WS upstream site. For the multispectral imagery captured in July 2018 at the treated WS study plot, the orthomosaic was 3.00 cm pixel^{-1}. The orthomosaic created from imagery captured in October of 2018 had a spatial resolution of 2.40 cm pixel^{-1}.

Image processing, including the creation of orthomosaics, was conducted using PhotoScan professional (Agisoft LLC, St Petersburg, Russia). Photo alignment was performed using the highest accuracy setting, with a key point limit of 120,000 and a tie point limit of 30,000. Adaptive camera fitting was performed during this step. To create the dense cloud, the ultra-high quality setting with aggressive depth filtering was used. A mesh was created using the 2.5D height field for surface data with a high face count, using the sparse cloud. The mosaic blending mode was used to build the texture. The dense cloud was used to build the tiled model and digital elevation model (DEM). Orthomosaics were subsequently produced in PhotoScan using the created DEM. Areas with poor image quality and without sufficient overlap (on the edges of the orthomosaic) were excluded from analysis. Image analysis was conducted using ArcGIS (version 10.6, Redlands, CA, USA). No radiometric correction was performed; therefore, brightness numbers were used for image analysis. Georeferencing and alignment of images was performed in ArcGIS using landscape features (e.g., gully intersection points) and selected ground control reference markers with known latitude and longitude positions that could be easily identified in imagery.

2.4. Comparative Analysis, Ground Data versus UAV Imagery

RGB imagery (red, green, and blue wavelengths) and multispectral imagery (red, green, blue, near-infrared, and red-edge wavelengths) were used to assess canopy cover and vegetation cover characteristics. Four vegetation indices were selected to assess canopy cover characteristics of the untreated study plots (Table 2). As RGB cameras are often more accessible and affordable compared to cameras that capture multispectral imagery, we assessed the effectiveness of using RGB imagery for measuring canopy cover (both mature juniper and juniper saplings), vegetation cover, and juniper identification.

Table 2. Vegetation indices selected to assess vegetation and ground cover characteristics of the study plots. Names refer to reflectance values for each band. The TGI uses the wavelength (λ) for the red, green, and blue bands in the calculation. NIR refers to near-infrared.

Method	Formula
Triangular Greenness Index (TGI) [46]	$-0.5[(\lambda red - \lambda blue)(Red - Green) - (\lambda red - \lambda green)(Red - Blue)]$
Optimized Soil Adjusted Vegetation Index (OSAVI) [33]	$(NIR - Red)/(NIR + Red + 0.16)$
Normalized Difference Vegetation Index (NDVI) [31]	$(NIR - Red)/(NIR + Red)$
Total Ratio Vegetation Index (TRVI) [34]	$4[(NIR - Red)/(NIR + Red + Green + Blue)]$

In order to measure canopy cover, the triangular greenness index (TGI) [46], which indicates chlorophyll content, was applied to the visual imagery at the untreated WS. Additionally, three vegetation indices (NDVI, OSAVI, and TRVI) that utilize multispectral imagery were used in estimating canopy cover at the untreated WS. Vegetation indices were calculated using the Raster Calculator function in ArcGIS, which created a raster of one band with these values.

Support vector machine (SVM) supervised classification was also used to assess canopy cover in the untreated WS. RGB imagery (red, green, and blue wavelengths) and multispectral imagery (red, green, blue, near-infrared, and red-edge wavelengths) were used for classification. Using the Training Sample Manager within ArcGIS, polygons were drawn around representative samples of juniper, bare soil, and woody debris. Training sites were selected from different areas of the image to represent the range of reflectance characteristics of each class. An initial image was created using the Interactive Supervised Classification function in order to determine how well the land cover was represented and to determine if more training samples were needed. Once each class was differentiated, the training samples were saved and used for classification. The Train Support Vector Machine Classifier tool in ArcGIS was used to create an Esri Classifier Definition (.ecd) file for each orthomosaic. Using the .ecd file, each the Classify Raster tool was used to perform classification on each raster. In order to minimize noise within the image, and remove small isolated clusters of pixels, the Majority Filter was then applied. The general supervised classification procedure used in this research is shown in Figure 2.

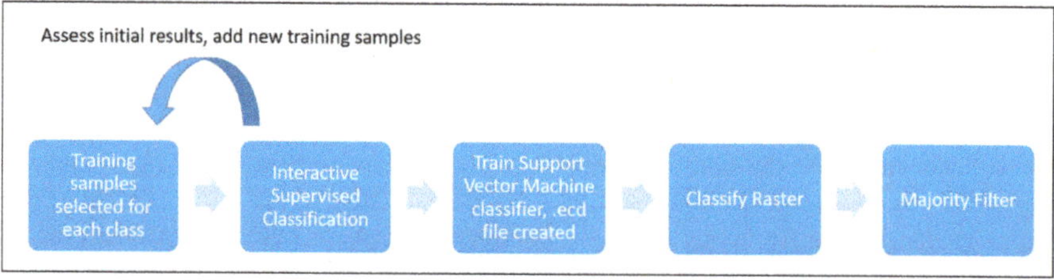

Figure 2. Supervised classification procedure performed in ArcMap.

A pixel-based analysis was conducted to assess juniper density and canopy cover, and total vegetation cover. By determining the number of pixels that correspond to vegetation compared to all other types of ground cover, the percentage of vegetation and canopy cover can be estimated within a given area. For the indices selected in this research, higher values (above 0) corresponded to greater photosynthetic activity or chlorophyll depending on the index applied. For instance, values for NDVI can range from −1 to 1, with areas of bare soil corresponding to values of approximately 0.025 or less, grasslands and shrub vegetation corresponding to values of around 0.09, and areas of dense vegetation corresponding to values of 0.4 or greater [47]. These values can vary depending on study site characteristics, vegetation type, season, sensor type and calibration, and weather conditions [48]. Based on visual inspection of the imagery (specifically examining values associated with bare ground, juniper canopy, woody debris, and shadows), threshold values were established for each index to separate vegetation from all other ground cover. The number of pixels with values greater than the threshold were divided by the total number of pixels in order calculate the percent of canopy cover or area covered by vegetation, similar to methods described by Wu [32].

Juniper identification, juniper sapling canopy cover, and vegetation ground cover in the treated WS were assessed for two dates, July 2018 and October 2018. Support vector machine supervised classification was applied to RGB imagery, multispectral imagery (red, green, blue, NIR, and red-edge bands) and imagery with multispectral bands and NDVI values at the treated WS. The same supervised classification process described above was used in the treated WS; however, training samples in the treated WS were divided into four main categories: juniper, other vegetation, woody debris, and bare ground.

The number of pixels classified as juniper, other vegetation, bare ground, or woody debris was tabulated. Similar to calculations made for canopy cover in the untreated plots, the number of pixels represented by each class was divided by the total number of pixels to determine a percent cover of each class in the treated WS. This was then compared to estimates of juniper sapling density and vegetated ground cover calculated using the belt transect method and to the results found using line-point intercept surveys conducted in 2018 at this study site [49].

To assess the accuracy of the supervised classification in the treated WS, 249 random points were selected in each orthomosaic using the ArcGIS random point tool (one random point corresponded to a board used as a ground control point and was subsequently excluded from this analysis). Additionally, 67 pixels corresponding to juniper saplings were identified within the image and used to assess the accuracy of juniper detection specifically. For accuracy analysis, four main classes were utilized: juniper, other vegetation, woody debris, and bare ground. No assessment was made of the accuracy of detecting specific vegetation species other than western juniper. A confusion matrix was created for each orthomosaic in the treated WS. From the confusion matrix, the user's accuracy (indication of Type I error), producer's accuracy (indication of Type II error), Cohen's kappa coefficient, and overall method accuracy were calculated.

3. Results

3.1. Ground-Based Vegetation Data Results

Tree Density, Height, and Canopy Cover

A greater number of juniper trees were observed in the untreated WS than in the treated WS. Based on ground counts in the two monitoring plots, average juniper tree density (of all age classes) was 797 trees ha^{-1} for the untreated WS. In the treated WS, juniper density was calculated to be 313 juniper saplings ha^{-1} based on the 41 belt transects distributed across the watershed. Mean sapling density was 473 trees ha^{-1} based on ground counts from the two 2000 m^2 monitoring plots at the valley and upstream locations.

The mean height of all juniper saplings surveyed in the treated WS (n = 113) was 0.75 m, ranging from 0.09 to 2.08 m. Tree crown width ranged from 0.09 to 1.4 m. Data from a subsample (n = 18) of saplings showed mean age tree was 9 years, ranging from 1 to 15 years. On average, saplings grew

0.1 m year^{-1} in height and 0.04 m year^{-1} in width. Mean sapling canopy cover was calculated to be 0.7% at the treated WS based on the belt transects.

In the untreated WS, adult trees (based on having a canopy diameter of 1.5 m or greater) made up 26% (34 of 133 trees) of all juniper at the valley monitoring plot and 16% (29 of 186 trees) of all juniper in the upstream monitoring plot. Mean sapling density was estimated to be 640 trees ha^{-1} at the untreated WS. Tree canopy cover was 30.4% for the valley plot and 28.0% for the upstream plot in the untreated WS.

3.2. UAV-Based Vegetation Data Results

3.2.1. Tree Canopy Cover at the Untreated Watershed (WS)

Canopy cover estimates using the UAV-based imagery at both untreated WS plots varied between vegetation indices that used visual or multispectral data (Table 3). The threshold value to determine vegetation was 0.05 for NDVI and OSAVI, 0.1 was used for TRVI and 0 was used for TGI. All pixels valued at and below the corresponding index threshold were considered to be non-vegetated areas. At the untreated WS upstream plot, NDVI, OSAVI and TRVI based estimates of canopy cover ranged from 26.1% to 27.3% (0.7% to 1.9% less than ground observations) while canopy cover measurements using TGI indicated 22.8% canopy cover (5.2% less than ground calculations). At the untreated WS valley plot, canopy cover estimates using NDVI, OSAVI, and TRVI were 33.7% to 34.5% (3.3% to 4.1% greater than ground measurements) (Figure 3). Canopy cover estimates using TGI at the untreated WS valley plot showed the largest difference from ground-based measurements at 21.2% (9.2% lower than ground estimates).

Table 3. Canopy cover at the untreated study plots. Method refers to the vegetation index used to calculate the canopy cover. TGI is calculated using reflectance values from the RGB (red, green, and blue bands) imagery. NDVI, OSAVI, and TRVI are calculated using reflectance values from the multispectral (red, green, blue, near-infrared, and red-edge bands) imagery. Values above the threshold value are considered vegetation. RGB classification refers to the support vector machine supervised classification performed using visual imagery. MS (multispectral bands: red, green, blue, near-infrared, and red-edge) classification refers to supervised classification performed using multispectral imagery. Ground refers to ground-based measurements of canopy cover made at each study plot.

Location	Method	Canopy Cover (%)	Threshold
Untreated WS Upstream	TGI	22.8	0
	NDVI	26.1	0.05
	OSAVI	26.7	0.05
	TRVI	27.3	0.1
	RGB classification	26.6	N/A
	MS classification	30.3	N/A
	Ground	28.0	N/A
Untreated WS Valley	TGI	21.2	0
	NDVI	33.7	0.05
	OSAVI	33.7	0.05
	TRVI	34.5	0.1
	RGB classification	29.5	N/A
	MS classification	29.4	N/A
	Ground	30.4	N/A

At the untreated WS valley site, estimates of canopy cover using SVM supervised classification (both RGB and multispectral imagery) were closer to ground-based results compared to the use of vegetation indices (Table 3). At the untreated WS upstream site, TRVI estimated canopy cover better than the SVM classification using RGB or multispectral imagery.

Figure 3. Untreated WS valley canopy cover. Darker shades correspond to lower vegetation index values and lighter shades correspond to higher values. NDVI, TRVI, and OSAVI values were similar, and therefore only OSAVI is shown for comparison. Differences can be seen in the characterization of canopy cover and in the shadows under the canopy between the OSAVI and TGI images.

3.2.2. Vegetation Cover and Juniper Sapling Density at the Treated WS

Based on the number of pixels corresponding to vegetation and non-vegetated areas, estimates of total vegetated ground cover based at the treated WS were similar between methods with the exception of the RGB imagery from October 2018 (Table 4). Results of these methods were also similar to line-point intercept surveys conducted in 2018 [49]. However, visual inspection of some areas of the classified rasters indicated regions where areas of bare ground and vegetation were misclassified. The use of NDVI with multispectral imagery resulted in a small difference in overall vegetated cover estimate for October imagery (0.1%) and a 1.9% difference in the estimate of vegetation cover for July imagery.

Table 4. Characterization of ground cover by pixel-based analysis from supervised classification from July and October 2018 at the Treated WS. RGB refers to supervised classification of RGB bands (red, green, and blue) only. Multispectral data ("MS only") uses reflectance values from the red, green, blue, near-infrared, and red-edge wavelengths. Multispectral with NDVI ("MS+NDVI") used the multispectral bands with the addition of NDVI values for classification. Non-vegetated ground cover refers to all other types of ground cover evaluated: bare ground and woody debris. Ground-based results are based on data collected from belt-transects and line-point transects from a study in 2018 [49].

	Juniper Cover (%)	Vegetation Cover (%)	Non-Vegetated Cover (%)
Jul 18: RGB	4.8	43.2	56.8
Jul 18: MS only	6.5	41.2	58.8
Jul 18: MS+NDVI	7.5	43.1	56.9
Oct 18: RGB	3.5	50.3	49.7
Oct 18: MS only	0.7	42.8	57.2
Oct 18: MS+NDVI	1.0	42.7	57.3
Ground	0.7	43.1	56.9

Based on the number of pixels classified as juniper, estimates of juniper sapling canopy cover using multispectral data (with and without NDVI values) from October 2018 were similar to ground-based estimates. Estimates of juniper sapling cover from multispectral imagery in July 2018 were five to six times those of the ground-based measurements. Juniper density estimates at the treated WS based on RGB imagery were 3.5% (October 2018) and 4.8% (July 2018), compared to the 0.7% juniper density calculated using belt transects.

Overall, identification of juniper saplings using supervised classification was more accurate in the October orthomosaic compared to that of imagery from July (Figure 4, Tables 5 and 6). User's accuracy of juniper was also greater in October compared to July, regardless of the inclusion of NDVI values or if multispectral or RGB imagery was used. User's accuracy ranged from 72.6% (RGB imagery from

July 2018) to 100% (October imagery without NDVI and October RGB imagery). Producer's accuracy for juniper ranged from 64.3% (July RGB imagery) to 88.6% (October imagery with and without NDVI). For both months, the use of NDVI resulted in slight differences in user's and producer's accuracy of juniper. The use of RGB imagery was also associated with somewhat lower producer's accuracies compared to multispectral imagery.

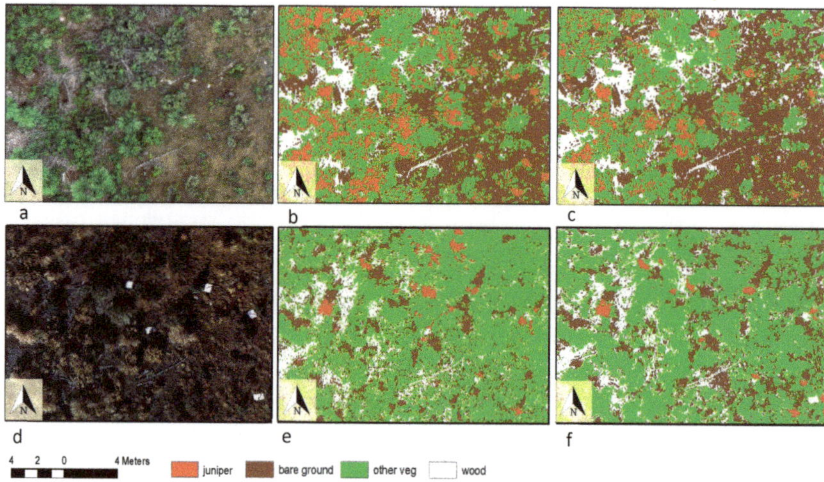

Figure 4. Subset of orthomosaic and classified rasters for the treated WS. July imagery is shown in top row: (**a**) subset of original orthomosaic, (**b**) classification of RGB (red, green, and blue wavelengths) raster, and (**c**) classification of MS (multispectral bands: red, green, blue, near-infrared, and red-edge wavelengths) without NDVI values. October imagery is displayed on the bottom row: (**d**) subset of original orthomosaic, (**e**) classification of RGB raster, and (**f**) classification of MS raster without NDVI. All classified images display results following the application of the Majority Filter tool. Red shading indicates pixels classified as juniper while pixels shaded green represent pixels identified as any other vegetation type. White cardboard was used to identify juniper in the October image (**d**) but is not present in the July image (**a**). "Other veg" refers to all vegetation not classified as juniper.

The accuracy of these methods to assess other types of ground cover (non-juniper vegetation, bare ground, and woody debris) was also compared (Tables 5 and 6). However, specific vegetation species other than juniper (such as sagebrush) were not analyzed for accuracy, and all pixels that corresponded to non-juniper vegetation were grouped together for analysis. Misclassification of pixels corresponding to areas of bare ground and woody debris occurred more frequently in both October orthomosaics compared to the July orthomosaics (Figure 4, Tables 5 and 6).

The overall accuracy of the supervised classification for all classes analyzed (juniper, bare ground, other vegetation, and woody debris) ranged from 76.6% (July RGB imagery) to 80.7% (July multispectral imagery with NDVI) (Table 7). The greater overall accuracy of the July orthomosaic can be largely attributed to the higher producer's accuracy of woody debris for this month. Values of Cohen's kappa coefficient for each method were similar, ranging from 0.70 to 0.74. The kappa coefficient for the juniper class only was 0.88 for both October orthomosaics, and 0.69 (multispectral imagery only) and 0.71 (multispectral imagery with NDVI) for the July orthomosaics. The use of RGB imagery resulted in very low kappa coefficients for the juniper class specifically: −0.08 for RGB imagery in July and 0.47 for RGB imagery in October.

Table 5. Confusion matrix for supervised classification for July 2018, for RGB (red, green, and blue wavelengths), multispectral imagery (red, green, blue, near-infrared, and red-edge wavelengths), and multispectral imagery with NDVI values. Reference pixels are displayed by column and classified pixels are displayed by row. All vegetation that was not juniper was grouped under the class of "Other Veg".

July RGB	Reference				
Classified	Juniper	Bare Ground	Other Veg	Wood	User's Accuracy
Juniper	45	2	15	0	72.6
Bare Ground	0	73	3	8	86.9
Other Veg	25	6	86	0	73.5
Wood	0	8	7	38	71.7
Producer's Accuracy	64.3	82.0	77.5	82.6	
July without NDVI	Reference				
Classified	Juniper	Bare Ground	Other Veg	Wood	User's Accuracy
Juniper	48	1	11	0	80.0
Bare Ground	0	67	3	1	94.4
Other Veg	22	6	89	0	76.1
Wood	0	15	8	45	66.2
Producer's Accuracy	68.6	75.3	80.2	97.8	
July NDVI	Reference				
Classified	Juniper	Bare Ground	Other Veg	Wood	User's Accuracy
Juniper	50	1	15	0	75.8
Bare Ground	0	73	2	2	94.8
Other Veg	20	6	88	0	77.2
Wood	0	9	6	44	74.6
Producer's Accuracy	71.4	82.0	79.3	95.7	

Table 6. Confusion matrix for supervised classification for October 2018, for RGB (red, green, and blue wavelengths), multispectral imagery (red, green, blue, near-infrared, and red-edge wavelengths), and multispectral imagery with NDVI values. Reference pixels are displayed by column and classified pixels are displayed by row. All vegetation that was not juniper was grouped under the class of "Other Veg".

Oct RGB	Reference				
Classified	Juniper	Bare Ground	Other Veg	Wood	User's Accuracy
Juniper	58	0	0	0	100.0
Bare Ground	0	67	9	5	82.7
Other Veg	11	17	76	7	68.5
Wood	1	16	7	42	63.6
Producer's Accuracy	82.9	67.0	82.6	77.8	
Oct without NDVI	Reference				
Classified	Juniper	Bare Ground	Other Veg	Wood	User's Accuracy
Juniper	62	0	0	0	100.0
Bare Ground	0	67	9	5	82.7
Other Veg	8	17	76	7	70.4
Wood	0	17	7	41	63.1
Producer's Accuracy	88.6	66.3	82.6	77.4	
Oct NDVI	Reference				
Classified	Juniper	Bare Ground	Other Veg	Wood	User's Accuracy
Juniper	62	1	0	0	98.4
Bare Ground	0	68	7	5	85.0
Other Veg	8	16	77	6	72.0
Wood	0	13	9	44	66.7
Producer's Accuracy	88.6	69.4	82.8	80.0	

Table 7. Overall accuracy and Cohen's kappa coefficient for supervised classification of each orthomosaic. Method refers to the kappa coefficient across all classes while juniper refers to the kappa coefficient only for juniper.

	Overall Accuracy (%)	Kappa (Method)	Kappa (Juniper Only)
Jul 18: RGB	76.6	0.72	−0.08
Jul 18: MS only	78.8	0.71	0.69
Jul 18: MS+NDVI	80.7	0.74	0.71
Oct 18:RGB	76.9	0.71	0.47
Oct 18: MS only	77.8	0.70	0.88
Oct 18: MS+NDVI	79.4	0.72	0.88

4. Discussion

This research examined the use of ground and UAV-based techniques to assess vegetation characteristics in two watersheds, one dominated by juniper and one where the majority of juniper was removed 14 years ago. We hypothesized that the use of high-resolution UAV imagery could be used to reasonably estimate canopy cover and juniper density at this study site.

4.1. Juniper Canopy Cover

Our ground-based tree canopy cover estimates (0.7% treated WS; 29.2% untreated WS) were similar to those by Ray et al. [44] who, 10 years post-treatment, estimated <1% juniper cover in the treated WS and 30% in the untreated. Our results are also similar to those by Bates et al. [50] for a similar western juniper ecosystem in southeastern Oregon; where juniper saplings occupied 0.8% of the treated plots and mature juniper occupied 29.6% of the control plots 12 years post-treatment.

Results also indicate that canopy cover estimated from vegetation indices (TRVI, NDVI, and OSAVI) derived from multispectral imagery were similar to the ground-based canopy cover estimates. Our results are similar to those by Davies et al. [51] who found a strong correlation between juniper cover estimates made using National Agriculture Imagery Program imagery and ground-based measurements using line intercepts.

Canopy cover estimates using a vegetation index (TGI) based on RGB data were not as close to our ground-based estimates when compared to those obtained using multispectral imagery. However, when supervised classification was used with RGB imagery, we found that canopy cover estimates substantially improved when compared to the use of TGI. This is similar to the findings from other studies that have successfully used RGB imagery to estimate canopy cover in other settings such as dense beech forests [26] and rice fields [52].

4.2. Juniper Sapling Density and Vegetative Cover

Our ground-based estimates of juniper density (797 trees ha^{-1}) in the untreated WS were similar to those (743 trees ha^{-1}) reported by Fisher [43] in 2004. Supervised classification applied to multispectral imagery (with and without NDVI) collected in October 2018 produced similar estimates of juniper sapling density (0.7% and 1.0%) compared to our ground-based results (0.7%).

Several studies [24,53] have highlighted the impact of seasonal collection on the accuracy of vegetation detection. In this study, the accuracy of juniper detection in the treated WS monitoring plot was greater for October imagery when compared to July imagery. Visible differences in vegetation were apparent in the imagery collected in the fall compared to the summer, this due in part to the non-juniper vegetation had senesced or displayed reduced vigor compared to the July imagery. While data collection occurred at around the middle of the day for both flights, differences in illumination were also clear in the images. The use of RGB imagery also resulted in lower producer's and user's accuracies of juniper identification compared to multispectral imagery of the same month. Furthermore, low kappa coefficients of the juniper class for RGB imagery were observed during both seasons while much higher kappa coefficients for the juniper class were observed with multispectral imagery in

October. Similar to our findings, Everitt et al. [54] found differences in reflectance characteristics between juniper and surrounding vegetation during summer and spring. Juniper in northwest Texas was also found to have different reflectance characteristics than other species during the month of February but no other times of the year [55].

4.3. Study Limitations and Future Research

The differences in canopy cover estimates and juniper density observed may be related to techniques and timing of data collection. While all flights were conducted around the same time of day, differences in cloud cover and topography may have influenced shading and vegetation index values. Small differences in threshold values will likely influence the canopy cover estimate and inherent differences in the ground-based measurement methods have also been found to contribute to differences in canopy cover estimates in semiarid woodlands [56]. The belt transects in the treated WS captured a wider range of topography and slopes within the watershed compared to the UAV-monitoring plot, which may not be representative of the differing vegetation characteristics.

By using small monitoring plots we were able to compare ground measurements more directly to UAV analysis to determine accuracy. However, a larger study area would encompass more topographical features of the watershed allowing us to compare results by aspect and slope. In addition to density and canopy cover, sapling height is an important characteristic in understanding juniper re-establishment. The use of UAV-based imagery to measure tree height has been demonstrated in several studies [57–59], and may provide important information regarding juniper height at this study site in future research. However, the height of juniper saplings may be similar to much of the surrounding vegetation (e.g., sagebrush) so consideration should be given to the age and structure of juniper stands.

The choice of pixel-based classification methods can influence results. A support vector machine (SVM) approach was used for supervised classification in this study. The use of SVM offers advantages over some other supervised classification methods, such as the maximum likelihood classifier, as it does not require the data to be normally distributed. However, Otukei and Blaschke [60] found that decision trees outperformed both support vector machine and maximum likelihood classifier approaches for land cover classification in open woodlands in Portugal, although all three methods produced acceptable accuracies. Another study, Joy et al. [61] successfully used decision trees to identify key vegetation types within a mixed woodland ecosystem that included pinyon-juniper species.

The accuracy of supervised classification is also dependent upon the training samples. Challenges associated with pixel-based analysis result when individual pixels may represent different classes (e.g., bare ground and wood), which may have accounted for some misclassification of wood and bare ground pixels in this study. If classes are very similar, misidentification and misclassification can occur. The use of a hybrid approach may also improve accuracy at our study site. Kumar et al. [62] found that the use of unsupervised and supervised classification together improved land cover classification in a semiarid region over using either approach alone. Additionally, in this study we used the pixel digital numbers, without radiometric calibration, for analysis. While imagery was visually assessed, in order to assess temporal changes or fuse imagery from multiple dates together, radiometric corrections should be made in future research.

This study utilized pixel-based image analysis based on pixel brightness values. Future research utilizing object-based image analysis (OBIA) incorporating shape and texture into classification may help delineate between juniper and other vegetation species. Baena et al. [63] found that OBIA, when used in combination with structure from motion (SfM) derived height models, could be used to assess the density of tree species in Northern Peru. While the ArcGIS Majority Filter tool was used in this research to minimize isolated pixels, the use of OBIA has been shown to reduce the amount of scattered pixels in high-resolution imagery [64].

The results of this study demonstrated the potential use of quadcopter UAV for evaluating juniper canopy cover and density, when seasonal limitations for data collection are considered. As in

Breckenridge et al. [65], our vegetative cover measurements using UAV-based multispectral imagery were similar to ground-based measurements. However, the accuracy of juniper sapling identification varied between seasons. Similar to the results found by Tay et al. [39], we found that pixel-based classification applied to UAV images can accurately detect and monitor vegetation. Given the large scale of juniper expansion and time requirements associated with ground surveys, the use of UAV offers the advantage of more efficient data collection compared to using ground-based techniques alone. UAV also offer a high-resolution, flexible platform which can be used by land managers to target specific study sites, times, and objectives.

Our study was specifically designed to evaluate UAV techniques for the management of juniper ecosystems. Similar to research by Lehmann et al. [66], future research at this study site includes expanding the techniques used in this study to characterize the spatial distribution of invasive species such as cheatgrass, which negatively affect rangeland ecosystems in the Pacific Northwest. A variety of other UAV applications such as in forest conservation planning, post-fire recovery, and estimation of dendrometric parameters for timber extraction forecast will likely become more common in the near future [67] because of the high cost, time and labor involved in traditional extensive field methods, especially in inaccessible locations [68].

5. Conclusions

This study evaluated western juniper canopy cover and density in a treated (juniper removed) and an untreated WS using ground-based and UAV-based methods. This research found that, as expected, juniper canopy cover and density were greater at the untreated WS compared to the treated WS. When supervised classification or multispectral vegetation indices were used, estimates of mature juniper canopy cover were similar to ground-based results. Additionally, we found that juniper sapling reestablishment post-treatment was of similar magnitude to that obtained in previous studies by different methods in the same ecosystem.

The results of this study also emphasize the importance of considering the seasonal characteristics of vegetation when collecting data. Juniper identification was more accurately achieved with October multispectral imagery than with July multispectral imagery or with RGB imagery. However, estimates of vegetation cover in the treated WS were similar, with the exception of RGB imagery from October, to ground-based estimates regardless of the season of collection. Although specific objectives and data collection regimes should be considered, UAV techniques are promising tools for monitoring western juniper expansion and monitoring vegetation cover in semiarid ecosystems.

Author Contributions: N.D. and C.G.O. developed the study design and conducted field data collection and analyses. R.M.-G. provided expert knowledge used in data analysis and interpretation. All co-authors contributed to the writing of the manuscript.

Funding: This study was funded in part by the Oregon Beef Council, Oregon Watershed Enhancement Board, USDA NIFA, and the Oregon Agricultural Experiment Station.

Acknowledgments: The authors gratefully acknowledge the continuous support of the Hatfield High Desert Ranch, the U.S. Department of Interior Bureau of Land Management—Prineville Office, and the OSU's Extension Service, in this research effort. Our thanks go to Tim Deboodt, Michael Fisher, and John Buckhouse for paving the road for this important long-term study. Also, we want to thank the multiple other graduate and undergraduate students from Oregon State University, and volunteers, who participated in various field data collection activities related to the results here presented.

Conflicts of Interest: The authors declare no conflict of interest.

References

1. Miller, R.F.; Tausch, R.J.; McAarthur, E.D.; Johnson, D.D.; Sanderson, S.C. *Age Structure and Expansion of Pinon-Juniper Woodlands: A Regional Perspective in the Intermountain West*; Research Paper RMRS-RP-69; U.S. Department of Agriculture, Forest Service: Fort Collins, CO, USA, 2008.
2. Miller, R.F.; Bates, J.D.; Svejcar, T.J.; Pierson, F.B.; Eddleman, L.E. *Biology, Ecology, and Management of Western Juniper (Juniperus occidentalis)*; Oregon State University, Agricultural Experiment Station: Corvallis, OR, USA, 2005; Volume 152.
3. Caracciolo, D.; Istanbulluoglu, E.; Noto, L.V. An ecohydrological cellular automata model investigation of juniper tree encroachment in a western North American landscape. *Ecosystems* **2017**, *20*, 1104–1123. [CrossRef]
4. Waichler, W.S.; Miller, R.F.; Doescher, P.S. Community characteristics of old-growth western juniper woodlands. *J. Range Manag.* **2001**, *54*, 518–527. [CrossRef]
5. Miller, R.F.; Rose, J.A. Historic expansion of Juniperus occidentalis (western juniper) in southeastern Oregon. *Gt. Basin Nat.* **1995**, *55*, 37–45.
6. Coultrap, D.E.; Fulgham, K.O.; Lancaster, D.L.; Gustafson, J.; Lile, D.F.; George, M.R. Relationships between western juniper (Juniperus occidentalis) and understory vegetation. *Invasive Plant Sci. Manag.* **2008**, *1*, 3–11. [CrossRef]
7. Miller, R.F.; Svejcar, T.J.; Rose, J.A. Impacts of western juniper on plant community composition and structure. *J. Range Manag.* **2000**, *53*, 574–585. [CrossRef]
8. Lebron, I.; Madsen, M.D.; Chandler, D.G.; Robinson, D.A.; Wendroth, O.; Belnap, J. Ecohydrological controls on soil moisture and hydraulic conductivity within a pinyon-juniper woodland. *Water Resour. Res.* **2007**, *43*, W08422. [CrossRef]
9. Mollnau, C.; Newton, M.; Stringham, T. Soil water dynamics and water use in a western juniper (Juniperus occidentalis) woodland. *J. Arid Environ.* **2014**, *102*, 117–126. [CrossRef]
10. Pierson, F.B.; Williams, C.J.; Hardegree, S.P.; Clark, P.E.; Kormos, P.R.; Al-Hamdan, O.Z. Hydrologic and erosion responses of sagebrush steppe following juniper encroachment, wildfire, and tree cutting. *Rangel. Ecol. Manag.* **2013**, *66*, 274–289. [CrossRef]
11. Dittel, J.W.; Sanchez, D.; Ellsworth, L.M.; Morozumi, C.N.; Mata-Gonzalez, R. Vegetation response to juniper reduction and grazing exclusion in sagebrush-steppe habitat in eastern Oregon. *Rangel. Ecol. Manag.* **2018**, *71*, 213–219. [CrossRef]
12. Reid, K.; Wilcox, B.; Breshears, D.; Macdonald, L. Runoff and erosion in a pinon-juniper woodland: Influence of vegetation patches. *Soil Sci. Soc. Am. J.* **1999**, *63*, 1869–1879. [CrossRef]
13. Pierson, F.B.; Bates, J.D.; Svejcar, T.J.; Hardegree, S.P. Runoff and erosion after cutting western juniper. *Rangel. Ecol. Manag.* **2007**, *60*, 285–292. [CrossRef]
14. Petersen, S.L.; Stringham, T.K. Infiltration, runoff, and sediment yield in response to western juniper encroachment in southeast Oregon. *Rangel. Ecol. Manag.* **2008**, *61*, 74–81. [CrossRef]
15. O'Connor, C.; Miller, R.; Bates, J.D. Vegetation response to western juniper slash treatments. *Environ. Manag.* **2013**, *52*, 553–566. [CrossRef] [PubMed]
16. Pierson, F.B.; Williams, C.J.; Kormos, P.R.; Hardegree, S.P.; Clark, P.E.; Rau, B.M. Hydrologic vulnerability of sagebrush steppe following pinyon and juniper encroachment. *Rangel. Ecol. Manag.* **2010**, *63*, 614–629. [CrossRef]
17. Sankey, T.T.; Glenn, N.; Ehinger, S.; Boehm, A.; Hardegree, S. Characterizing western juniper expansion via a fusion of Landsat 5 Thematic mapper and lidar data. *Rangel. Ecol. Manag.* **2010**, *63*, 514–523. [CrossRef]
18. Petersen, S.L.; Stringham, T.K. Development of GIS-based models to predict plant community structure in relation to western juniper establishment. *For. Ecol. Manag.* **2008**, *256*, 981–989. [CrossRef]
19. Roundy, D.B.; Hulet, A.; Roundy, B.A.; Jensen, R.R.; Hinkle, J.B.; Crook, L.; Petersen, S.L. Estimating pinyon and juniper cover across Utah using NAIP imagery. *Environment* **2016**, *3*, 765–777. [CrossRef]
20. Yang, J.; Weisberg, P.J.; Bristow, N.A. Landsat remote sensing approaches for monitoring long-term tree cover dynamics in semi-arid woodlands: Comparison of vegetation indices and spectral mixture analysis. *Remote Sens. Environ.* **2012**, *119*, 62–71. [CrossRef]

21. Meddens, A.J.H.; Hicke, J.A.; Jacobs, B.F. Characterizing the response of piñon-juniper woodlands to mechanical restoration using high-resolution satellite imagery. *Rangel. Ecol. Manag.* **2016**, *69*, 215–223. [CrossRef]

22. Xian, G.; Homer, C.; Aldridge, C. Assessing long-term variations in sagebrush habitat—Characterization of spatial extents and distribution patterns using multi-temporal satellite remote-sensing data. *Int. J. Remote Sens.* **2012**, *33*, 2034–2058. [CrossRef]

23. Howell, R.G.; Petersen, S.L. A comparison of change detection measurements using object-based and pixel-based classification methods on western juniper dominated woodlands in eastern Oregon. *AIMS Environ. Sci.* **2017**, *4*, 348–357. [CrossRef]

24. Lu, B.; He, Y. Species classification using Unmanned Aerial Vehicle (UAV)-acquired high spatial resolution imagery in a heterogeneous grassland. *ISPRS J. Photogramm. Remote Sens.* **2017**, *128*, 73–85. [CrossRef]

25. D'Oleire-Oltmanns, S.; Marzolff, I.; Peter, K.D.; Ries, J.B. Unmanned aerial vehicle (UAV) for monitoring soil erosion in Morocco. *Remote Sens.* **2012**, *4*, 3390–3416. [CrossRef]

26. Chianucci, F.; Disperati, L.; Guzzi, D.; Bianchini, D.; Nardino, V.; Lastri, C.; Rindinella, A.; Corona, P. Estimation of canopy attributes in beech forests using true colour digital images from a small fixed-wing UAV. *Int. J. Appl. Earth Obs. Geoinform.* **2016**, *47*, 60–68. [CrossRef]

27. Carreiras, J.M.B.; Pereira, J.M.C.; Pereira, J.S. Estimation of tree canopy cover in evergreen oak woodlands using remote sensing. *For. Ecol. Manag.* **2006**, *223*, 45–53. [CrossRef]

28. Krofcheck, D.J.; Eitel, J.U.H.; Lippitt, C.D.; Vierling, L.A.; Schulthess, U.; Litvak, M.E. Remote sensing based simple models of GPP in both disturbed and undisturbed piñon-juniper woodlands in the southwestern U.S. *Remote Sens.* **2016**, *8*, 20. [CrossRef]

29. Zhou, X.; Zheng, H.B.; Xu, X.Q.; He, J.Y.; Ge, X.K.; Yao, X.; Cheng, T.; Zhu, Y.; Cao, W.X.; Tian, Y.C. Predicting grain yield in rice using multi-temporal vegetation indices from UAV-based multispectral and digital imagery. *ISPRS J. Photogramm. Remote Sens.* **2017**, *130*, 246–255. [CrossRef]

30. Akar, A.; Gökalp, E.; Akar, Ö.; Yilmaz, V. Improving classification accuracy of spectrally similar land covers in the rangeland and plateau areas with a combination of worldview-2 and UAV images. *Geocarto Int.* **2017**, *32*, 990–1003. [CrossRef]

31. Rouse, J.W.; Harlan, J.C.; Haas, R.H.; Schell, J.A.; Deering, D.W. *Monitoring the Vernal Advancement and Retrogradiation (Green Wave Effect) of Natural Vegetation*; Texas A&M University, Remote Sensing Center: College Station, TX, USA, 1974.

32. Wu, W. Derivation of tree canopy cover by multiscale remote sensing approach. *Int. Arch. Photogramm.* **2011**, 142–149. [CrossRef]

33. Rondeaux, G.; Steven, M.; Baret, F. Optimization of soil-adjusted vegetation indices. *Remote Sens. Environ.* **1996**, *55*, 95–107. [CrossRef]

34. Fadaei, H.; Suzuki, R.; Sakaic, T.; Toriid, K. A Proposed new vegetation index, the total ratio vegetation index (TRVI), for arid and semiarid regions. *ISPRS Int. Arch. Photogramm. Remote Sens. Spat. Inf. Sci.* **2012**, *1*, 403–407. [CrossRef]

35. Akkermans, T.; Van Rompaey, A.; Van Lipzig, N.; Moonen, P.; Verbist, B. Quantifying successional land cover after clearing of tropical rainforest along forest frontiers in the Congo Basin. *Phys. Geogr.* **2013**, *34*, 417–440. [CrossRef]

36. Koppad, A.G.; Janagoudar, B.S. Vegetation analysis and land use land cover classification of forest in Uttara Kannada District India using remote sensing and GIS techniques. *ISPRS Int. Arch. Photogramm. Remote Sens. Spat. Inf. Sci.* **2017**, *XLII-4/W5c*, 121–125. [CrossRef]

37. García-Romero, L.; Hernández-Cordero, A.I.; Hernández-Calvento, L.; Espino, E.P.; López-Valcarcel, B.G. Procedure to automate the classification and mapping of the vegetation density in arid aeolian sedimentary systems. *Prog. Phys. Geogr. Earth Environ.* **2018**, *42*, 330–351. [CrossRef]

38. Papadopoulos, A.V.; Kati, V.; Chachalis, D.; Kotoulas, V.; Stamatiadis, S. Weed mapping in cotton using ground-based sensors and GIS. *Environ. Monit. Assess.* **2018**, *190*, 622. [CrossRef]

39. Tay, J.Y.L.; Erfmeier, A.; Kalwij, J.M. Reaching new heights: Can drones replace current methods to study plant population dynamics? *Plant Ecol.* **2018**, *219*, 1139–1150. [CrossRef]

40. Bauer, T.; Strauss, P. A rule-based image analysis approach for calculating residues and vegetation cover under field conditions. *Catena* **2014**, *113*, 363–369. [CrossRef]

41. Ochoa, C.G.; Caruso, P.; Ray, G.; Deboodt, T.; Jarvis, W.T.; Guldan, S.J. Ecohydrologic connections in semiarid watershed systems of central Oregon USA. *Water* **2018**, *10*, 181. [CrossRef]
42. Deboodt, T.L. Watershed Response to Western Juniper Control. Ph.D. Dissertation, Oregon State University, Corvallis, OR, USA, 2008.
43. Fisher, M. Analysis of Hydrology and Erosion in Small, Paired Watersheds in a Juniper-Sagebrush Area of Central Oregon. Ph.D. Dissertation, Oregon State University, Corvallis, OR, USA, 2004.
44. Ray, G.; Ochoa, C.G.; Deboodt, T.; Mata-Gonzalez, R. Overstory–understory vegetation cover and soil water content observations in western juniper woodlands: A paired watershed study in Central Oregon, USA. *Forests* **2019**, *10*, 151. [CrossRef]
45. Phipps, R.L. *Collecting, Preparing, Crossdating, and Measuring Tree Increment Cores*; US Department of the Interior, Geological Survey: Washington, DC, USA, 1985.
46. Hunt, E.R.; Doraiswamy, P.C.; McMurtrey, J.E.; Daughtry, C.S.T.; Perry, E.M.; Akhmedov, B. A visible band index for remote sensing leaf chlorophyll content at the canopy scale. *Int. J. Appl. Earth Obs. Geoinform.* **2013**, *21*, 103–112. [CrossRef]
47. Holben, B.N. Characteristics of maximum-value composite images from temporal AVHRR data. *Int. J. Remote Sens.* **1986**, *7*, 1417–1434. [CrossRef]
48. Wang, R.; Gamon, J.A.; Montgomery, R.A.; Townsend, P.A.; Zygielbaum, A.I.; Bitan, K.; Tilman, D.; Cavender-Bares, J. Seasonal variation in the NDVI–species richness relationship in a prairie grassland experiment (Cedar Creek). *Remote Sens.* **2016**, *8*, 128. [CrossRef]
49. Durfee, N. Ecohydrologic Connections in Semiarid Rangeland Ecosystems in Oregon. Master's Thesis, Oregon State University, Corvallis, OR, USA, 2018.
50. Bates, J.D.; Svejcar, T.; Miller, R.; Davies, K.W. Plant community dynamics 25 years after juniper control. *Rangel. Ecol. Manag.* **2017**, *70*, 356–362. [CrossRef]
51. Davies, K.W.; Petersen, S.L.; Johnson, D.D.; Davis, D.B.; Madsen, M.D.; Zvirzdin, D.L.; Bates, J.D. Estimating juniper cover From National Agriculture Imagery Program (NAIP) imagery and evaluating relationships between potential cover and environmental variables. *Rangel. Ecol. Manag. Lawrence* **2010**, *63*, 630–637. [CrossRef]
52. Lee, K.J.; Lee, B.W. Estimating canopy cover from color digital camera image of rice field. *J. Crop Sci. Biotechnol.* **2011**, *14*, 151–155. [CrossRef]
53. Ahmadpour, A.; Shokri, M.; Solaimani, K.; Ghorbani, J. Evaluation of satellite data efficiency in identification of plant groups. *Acta Ecol. Sin.* **2011**, *31*, 303–309. [CrossRef]
54. Everitt, J.H.; Yang, C.; Johnson, H.B. Canopy spectra and remote sensing of ashe juniper and associated vegetation. *Environ. Monit. Assess.* **2007**, *130*, 403–413. [CrossRef]
55. Everitt, J.H.; Yang, C.; Racher, B.J.; Britton, C.M.; Davis, M.R. Remote sensing of redberry juniper in the Texas rolling plains. *J. Range Manag.* **2001**, *54*, 254–259. [CrossRef]
56. Ko, D.; Bristow, N.; Greenwood, D.; Weisberg, P. Canopy cover estimation in semiarid woodlands: Comparison of field-based and remote sensing methods. *For. Sci.* **2009**, *55*, 10.
57. Birdal, A.C.; Avdan, U.; Türk, T. Estimating tree heights with images from an unmanned aerial vehicle. *Geomat. Nat. Hazards Risk* **2017**, *8*, 1144–1156. [CrossRef]
58. Wallace, L.; Lucieer, A.; Malenovsky, Z.; Turner, D.; Vopenka, P. Assessment of forest structure using two UAV techniques: A comparison of airborne Laser scanning and structure from motion (sfm) point clouds. *Forests* **2016**, *7*, 62. [CrossRef]
59. Panagiotidis, D.; Abdollahnejad, A.; Surový, P.; Chiteculo, V. Determining tree height and crown diameter from high-resolution UAV imagery. *Int. J. Remote Sens.* **2017**, *38*, 2392–2410. [CrossRef]
60. Otukei, J.R.; Blaschke, T. Land cover change assessment using decision trees, support vector machines and maximum likelihood classification algorithms. *Int. J. Appl. Earth Obs. Geoinf.* **2010**, *12*, S27–S31. [CrossRef]
61. Joy, S.M.; Reich, R.M.; Reynolds, R.T. A non-parametric, supervised classification of vegetation types on the Kaibab National Forest using decision trees. *Int. J. Remote Sens.* **2003**, *24*, 1835–1852. [CrossRef]
62. Kumar, P.; Singh, B.K.; Rani, M. An efficient hybrid classification Approach for land use/land cover analysis in a semi-desert area using ETM+ and LISS-III sensor. *IEEE Sens. J.* **2013**, *13*, 2161–2165. [CrossRef]
63. Baena, S.; Moat, J.; Whaley, O.; Boyd, D.S. Identifying species from the air: UAVs and the very high resolution challenge for plant conservation. *PLoS ONE* **2017**, *12*, e0188714. [CrossRef]

64. Laliberte, A.S.; Rango, A.; Herrick, J.E.; Fredrickson, E.L.; Burkett, L. An object-based image analysis approach for determining fractional cover of senescent and green vegetation with digital plot photography. *J. Arid Environ.* **2007**, *69*, 1–14. [CrossRef]
65. Breckenridge, R.P.; Dakins, M.; Bunting, S.; Harbour, J.L.; Lee, R.D. Using unmanned helicopters to assess vegetation cover in sagebrush steppe ecosystems. *Rangel. Ecol. Manag.* **2012**, *65*, 362–370. [CrossRef]
66. Lehmann, J.R.K.; Prinz, T.; Ziller, S.R.; Thiele, J.; Heringer, G.; Meira-Neto, J.A.A.; Buttschardt, T.K. Open-source processing and analysis of aerial imagery acquired with a low-cost unmanned aerial system to support invasive plant management. *Front. Environ. Sci.* **2017**, *5*, 44. [CrossRef]
67. Torresan, C.; Berton, A.; Carotenuto, F.; Gennaro, S.F.D.; Gioli, B.; Matese, A.; Miglietta, F.; Vagnoli, C.; Zaldei, A.; Wallace, L. Forestry applications of UAVs in Europe: A review. *Int. J. Remote Sens.* **2017**, *38*, 2427–2447. [CrossRef]
68. Martínez-Salvador, M.; Mata-Gonzalez, R.; Pinedo-Alvarez, A.; Morales-Nieto, C.R.; Prieto-Amparán, J.A.; Vázquez-Quintero, G.; Villarreal-Guerrero, F. A spatial forestry productivity potential model for pinus arizonica engelm, a key timber species from Northwest Mexico. *Sustainability* **2019**, *11*, 829. [CrossRef]

Article

Robinia pseudoacacia L. in Short Rotation Coppice: Seed and Stump Shoot Reproduction as well as UAS-based Spreading Analysis

Christin Carl [1,2,*], Jan R. K. Lehmann [3], Dirk Landgraf [2] and Hans Pretzsch [1]

[1] Forest Growth and Yield Science, School of Life Sciences Weihenstephan, Technische Universität München, Hans-Carl-von-Carlowitz-Platz 2, 85354 Freising, Germany; Hans.Pretzsch@lrz.tum.de

[2] Forestry and Ecosystem Management, University of Applied Sciences Erfurt, Leipziger Strasse 77, 99085 Erfurt, Germany; dirk.landgraf@fh-erfurt.de

[3] Institute of Landscape Ecology, University of Muenster, Heisenbergstrasse 2, 48149 Muenster, Germany; jan.lehmann@uni-muenster.de

* Correspondence: christin.carl@tum.de; Tel.: +49-163-882-8583

Received: 23 January 2019; Accepted: 28 February 2019; Published: 6 March 2019

Abstract: Varying reproduction strategies are an important trait that tree species need in order both to survive and to spread. Black locust is able to reproduce via seeds, stump shoots, and root suckers. However, little research has been conducted on the reproduction and spreading of black locust in short rotation coppices. This research study focused on seed germination, stump shoot resprout, and spreading by root suckering of black locust in ten short rotation coppices in Germany. Seed experiments and sample plots were analyzed for the study. Spreading was detected and measured with unmanned aerial system (UAS)-based images and classification technology—object-based image analysis (OBIA). Additionally, the classification of single UAS images was tested by applying a convolutional neural network (CNN), a deep learning model. The analyses showed that seed germination increases with increasing warm-cold variety and scarification. Moreover, it was found that the number of shoots per stump decreases as shoot age increases. Furthermore, spreading increases with greater light availability and decreasing tillage. The OBIA and CNN image analysis technologies achieved 97% and 99.5% accuracy for black locust classification in UAS images. All in all, the three reproduction strategies of black locust in short rotation coppices differ with regards to initialization, intensity, and growth performance, but all play a role in the survival and spreading of black locust.

Keywords: *Robinia pseudoacacia* L.; reproduction; spreading; short rotation coppice; unmanned aerial system (UAS); object-based image analysis (OBIA); convolutional neural network (CNN)

---

1. Introduction

The spreading of tree species is influenced by overcoming barriers (e.g., geographical) as well as survival and reproduction strategies. Trees are able to reproduce generatively via seeds, and vegetatively as well, for example via stump shoots or root suckers. Nevertheless, little research has been conducted on the reproduction and spreading strategies of black locust in short rotation coppices. However, examining the details within a holistic perspective of reproduction strategies for tree species may improve the estimation of spreading and survival potential, especially of non-native or invasive tree species.

Black locust (*Robinia pseudoacacia* L.) originated in the eastern part of North America, particularly in the Appalachian regions [1,2]. By the early 17th century, black locust had been introduced to Europe [3–6]. Today, black locust appears in many European countries, especially in Hungary, Northern Germany, Western Poland, Czech Republic, Southern Slovakia, and Eastern Austria [6]. In Germany, a change in energy policies aimed towards a reduction in the use of fossil fuels and greater utilization of renewable energy since 2004 [7] marked an increase in fast-growing tree production, including black locust in short rotation coppices (harvested every 2–20 years) [8,9]. Nevertheless, some studies have declared black locust to be an invasive non-native tree species in Europe [6,10–12]. However, other studies consider black locust to be an alternative tree species, for example for European ash (*Fraxinus excelsior*) [13,14], whose high dieback is caused by the fungal pathogen *Hymenoscyphus fraxineus* [14].

Black locust possesses many growth characteristics that makes it ideal for short-rotation biomass production such as: rapid growth, high drought tolerance, and nitrogen-fixation [15,16], as well as the ability to reproduce via stump shoots in response to harvest. Black locust starts flowering at the age of six years [17]. Many insects benefit from this characteristic, including honeybees [18]. Moreover, the production of black locust honey is very common and economically important, especially in Hungary [17]. A maximum of 15,559 flowers per tree were counted in an eight-year-old black locust plantation [18], and 12,000 seeds/m^2 were identified in a monodominant stand [6,19]; black locust thus produces an abundance of seeds. As black locust belongs to the *Fabaceae* family, the seed coat is hard and impermeable [20,21]. Hence, seeds require scarification for successful germination [21–23] and priming of seeds is favorable [21,24,25]. Seeds prefer mineral-rich sandy and loamy-sandy soil [17,26,27]. Additionally, seeds are dormant [28] and Voss and Edward [29] observed germination of black locust seeds 88 years after the seeds were collected. Moreover, increasing germination and seedling density was observed in the year immediately following a fire event; it is likely that some seeds survive fire and benefit from the light availability [30–32]. Nevertheless, scarification intensity has not been researched much, particularly in seeds from an eight-year-old short rotation coppice.

Vegetative reproduction strategies of black locust include re-sprouting stump shoots and root suckers. Stump shoot sprouting occurs in many broad-leafed trees in response to harvest or damage and is a characteristic feature of the special forest management practice known as (short rotation) coppicing. Stump shoots typically grow in densely populated stands and often have several shoots per stump. The biomass production differs compared to non-harvested trees, especially for black locust [33]. To account for this, specific yield tables and growth models for coppices of different broad-leafing tree species were developed [34–46]. Moreover, in nutrient-poor and sandy forest stands, a decreasing number of black locust stump shoots with increasing shoot age and a wide variety of stump shoots depending on the stump distance and stump age was observed [47,48]. However, the sprouting intensity and dieback of black locust shoots with increasing shoot age in densely populated short rotation coppices have not been a focus of research up to now.

The spreading of black locust short rotation coppice seems to take place primarily via root suckering [49]. Root suckers are shoots which grow from adventitious buds in roots of trees or shrubs. Crosti et al. [50] observed decreasing spreading intensity of black locust in orchards and nearby forests in Italy. Furthermore, Vítková et al. [6] published an overview of the distribution of black locust in central and eastern Europe on a landscape scale based on inventory data. However, there is a lack of knowledge about the spreading potential of black locust on a small scale, particularly with respect to short rotation coppice. Thereby, tree species detection and spreading could be analyzed via field measurements [50], satellite data [51,52], airplanes, or unmanned aerial systems (UAS) [53–63]. Field measurements are usually time consuming. Satellite data can be used to detect ecosystem structures and changes for large areas [64–68]. However, fine-scale field maps are difficult to generate from satellite or airplane data because they have a low spatial and temporal resolution and are generally expensive. Therefore, UAS allow the detection of ecosystem structures and changes offering a higher spatial resolution of small, specific, and detailed vegetation structures [18,69]. Additionally, UAS have quick turnaround times, are very cost-efficient and are useful supplements to data from satellites, airplanes, terrestrial manual, and other data analyses. Invasiveness analysis using UAS was demonstrated by Lehmann et al. [70] for invasive *Acacia mangium* management in Brazil. Müllerová et al. [71] conclude that the near-infrared (NIR) spectrum is very important for detecting black locust via satellite and UAS. Common approaches to analyzing UAS images are pixel-based analysis [18] and object-based image analysis (OBIA) [70], primarily including structure from motion (SfM) point clouds [53–55,57,58,60]. Nevertheless, OBIA has not yet been used to analyze the spreading of black locust in short rotation coppices.

Furthermore, there is an increasing interest in machine learning algorithms for data and image analysis, such as the application of the random forest model [52,72–76], support vector machine [73–76], and deep learning algorithms, especially convolutional neural networks (CNNs) [62,73,75,77–79]. However, CNNs were not previously utilized for the classification of black locust in short rotation coppices under varying conditions in single images.

The main purpose of this study is to estimate and monitor the invasiveness potential of black locust. Therefore, in the current study, we analyze the seed germination, stump shoot survival, and spreading of black locust planted in short rotation coppices. We investigate the seed germination of 3000 seeds, focusing on six different seed experiments. After the treatment, the seeds were seeded in planting boxes and watered. Moreover, we estimate black locust stump shoot sprouting and dieback with the aid of sample plots in seven different sample areas, including 5244 stump shoots. For this purpose, the relationship of the planted trees and the current shoots per stump were calculated. Additionally, spreading was estimated in short rotation coppices at a length of 2124 m based on UAS images by using OBIA. Furthermore, a deep learning algorithm was tested to classify black locust under varying light conditions, flying altitudes, UAS, and cameras in single images. This study deals with the following research questions:

(1) What is the average germination of black locust seeds after six different seed treatments?
(2) How many stump shoots survive in short rotation coppices depending on shoot age?
(3) What is the average sprouting distance of black locust, depending on neighboring forest, meadow, farmland, and along a dirt road analyzed via OBIA in UAS images?
(4) What is the accuracy and loss of black locust classification in single UAS images under varying conditions by using a CNN?

2. Materials and Methods

2.1. Site Description

The ten analyzed study sites were located in northeastern Germany (Table 1, Figure 1). The annual precipitation ranges between 495 mm and 671 mm and the mean annual temperature ranges between 7.4 °C and 9.4 °C [80]. The elevation above sea level is 22 to 149 m. Seeds were collected in winter 2016 in two eight-year-old *R. pseudoacacia* short rotation plantations in Germany (Lauchhammer and Röblingen). Thereafter, the seeds were stored in a seed bank (Thuringia, Germany). Stump shoot sample plots in seven study sites were analyzed between November 2016 and March 2017. The UAS images and field data for the spreading analysis were collected in May 2018 in Grunow-Dammendorf (Germany).

Table 1. Site description including location, longitude (long), latitude (lat), tree/shoot age, and analysis methodology.

Site Abbreviation	Location	Long (°E)	Lat (°N)	Tree/Shoot Age	Analysis
BG	Blumberg	14°10′24″	53°12′25″	3	Stump shoot
BH	Buchholz	12°38′9″	53°15′34″	2	Stump shoot
DA	Grunow-Dammendorf	14°25′5″	52°8′26″	4	Spreading
GU	Gumtow	12°14′11″	52°59′46″	2	Stump shoot
KL	Klein Loitz	14°30′57″	51°36′35″	3	Stump shoot
LH	Lauchhammer	13°50′57″	51°32′20″	8	Seed collection
PA	Paulinenaue	12°43′52″	52°39′44″	3	Stump shoot
RM	Röblingen	11°42′42″	51°25′57″	8	Seed collection
WA	Wainsdorf	13°29′4″	51°24′50″	1	Stump shoot
WZ	Welzow	14°14′7″	51°33′32″	1	Stump shoot

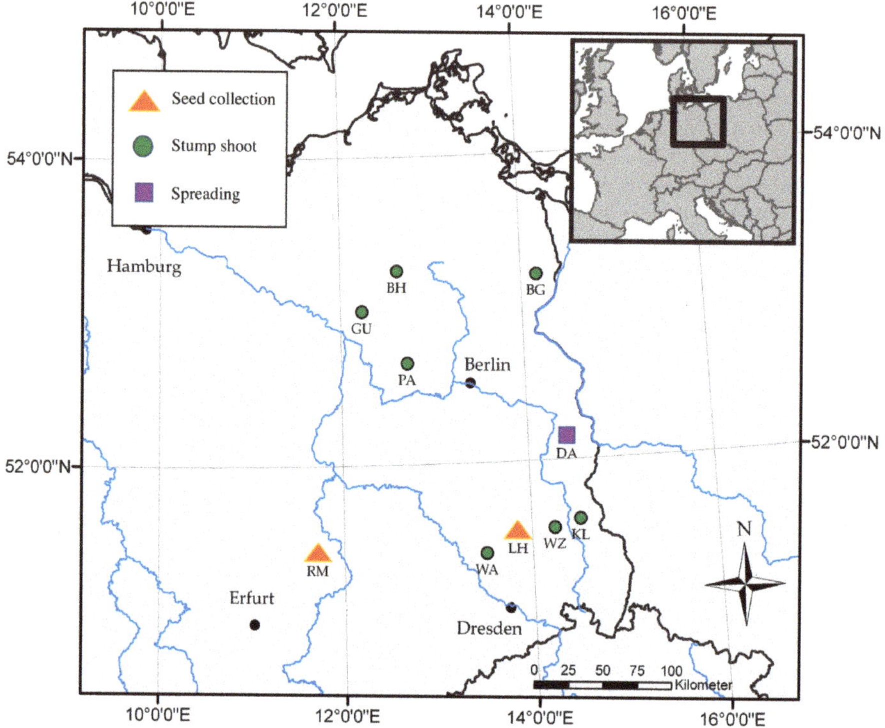

Figure 1. Map of the study area in Europe, indicating the ten *R. pseudoacacia* sampling sites in Germany [81]. Red triangles are sites visited for seed collection, green dots are the sites where sample plots for the stump shoot analysis were measured, and the purple square is the study site of the UAS image analysis, which was applied for the spreading detection. For an overview of site abbreviations, see Table 1.

2.2. Reproduction Analysis

The analysis was divided into three parts (Figure 2) to estimate the reproduction and spreading of black locust: seed, stump shoot, and root suckering analysis. Seed germination was undertaken to determine differences in seed treatments. Sample plots were used for the stump shoot analysis, which was conducted to analyze the number of shoots per stump depending on the shoot age. Spreading was analyzed by using an UAS and image analysis to observe the spreading from a bird's-eye view.

2.3. Seed Germination

The analysis of the seeds started in March 2018 and finished in July 2018. The seeds were seeded in sandy mineral soil in planting boxes. We tested 100 seeds in five iterations for six treatments (3000 seeds):

(I)   seeds were seeded and watered,

(II)   seeds were soaked for 24 h in water at a water temperature of 18 °C (64.4 °F) and then seeded and watered,

(III) seeds were stored at an air temperature of 45 °C (113 °F) for two hours, and thereafter stored at an air temperature of −20 °C (−4 °F) for a further two hours and then seeded and watered,

(IV)   seeds were stored at an air temperature of 60 °C (140 °F) for two hours and thereafter stored for two hours at an air temperature of −20 °C (−4 °F) and then seeded and watered,

(V)   seeds were scalded with hot water, seeded and watered, and

(VI)   seeds were mechanically scarified, seeded and watered.

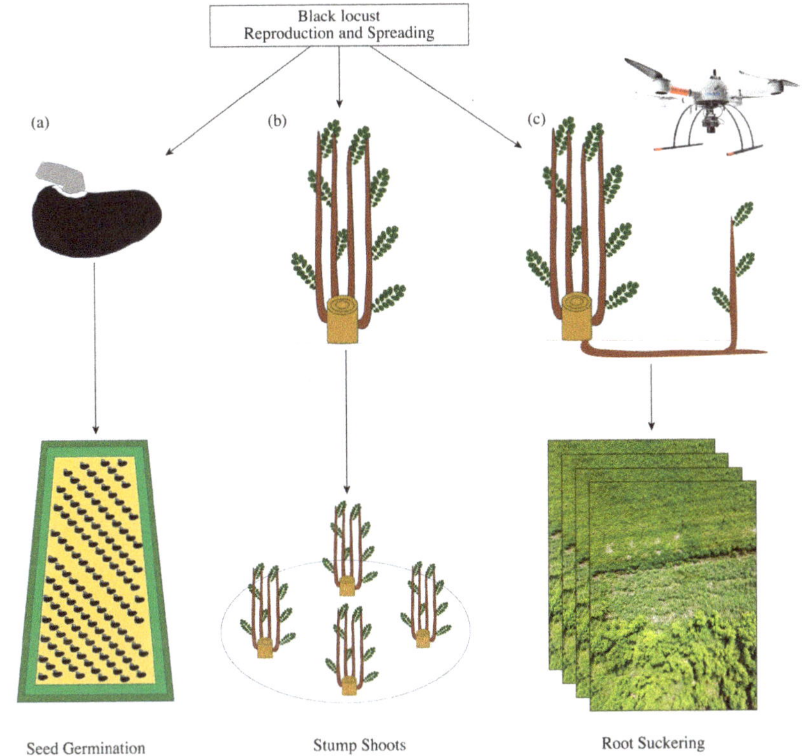

Figure 2. Flow chart showing the reproduction and spreading analysis: seed germination (**a**), stump shoot analysis (**b**), and root suckering (spreading) analysis (**c**). Seeds were seeded in planting boxes, and watered. Stump shoots were analyzed in sample plots. Spreading was estimated with the aid of UAS-images [82].

We calculated the average seed germination, standard deviation (SD) and standard error of the mean (SE) for each seed treatment. Furthermore, we performed an analysis of variance (ANOVA) with the software R [83]. Seed data were (1) variance heterogeneous and (2) normally distributed. Therefore, we applied a one-way analysis of means and pairwise *t*-test for pairwise comparisons [83,84].

2.4. Stump Shoot Analysis

In total, 5244 black locust stump shoots were investigated, measuring leafless stump shoots in 33 sample plots. Sample plots sizes ranged from 4 to 6 m radius and contained a minimum of 150 stump shoots. All stump shoots in the sample plot were counted. Furthermore, the planting distance was listed by the land manager and checked on the study sites. The measured number of stump shoots per sample plot was divided by the number of established/planted black locust trees to calculate the average number of shoots per stump, per site, and per age class. Moreover, we calculated SD, SE, confidence interval (CI) with a default of 95% for each age class and performed an ANOVA [83]. Stump shoot data were (1) variance homogeneous and (2) normally distributed. Hence, we applied an ANOVA as global test and Tukey test for pairwise comparisons [83,84].

2.5. Root Suckering (Spreading)

Two quadrocopters were used as a UAS platform: Microdrones MD4-1000 [82] and dji Mavic Pro [85]. The Microdrones MD4-1000 was equipped with two camera systems synchronously for the image collection: a multispectral MAPIR Survey 3 (red, green, NIR) camera [86] with an image size of 4000 × 3000 and a red-green-blue (RGB) SONY-ILCE-5100 [87] camera with a picture size of 6000 × 4000. Furthermore, the dji Mavic Pro took images with the aid of the RGB camera DJI FC220 (picture size of 4000 × 3000) [85]. Weather conditions were calm, and the lighting conditions varied between sunny and cloudy.

2.5.1. Object-Based Analysis (OBIA)

The analyzed total length was 2124 m, whereby the neighboring areas were comprised of meadow (726 m), farmland (565 m), dirt road (293 m), and pine forest (540 m). For georeferencing purposes, 13 white ground control points were randomly placed. In addition, planting lines were calibrated by using a mobile differential global positioning system [88]. The altitude above ground level was set at 30 m. To create 3D point cloud surface models, the image side and forward overlap were set to 80% [89]. RGB and NIR image data were orthorectified and mosaicked using Pix4D mapper software [90] to create a high-resolution orthoimagery. An SfM approach was used to calculate the digital surface models.

In eCognition Developer software [91], the classification procedure via OBIA consisted of two major steps: (A) segmentation and (B) classification. The multiresolution segmentation was used to aggregate neighboring pixels into segments based on homogeneity criteria (shape, texture, color, compactness, smoothness) and a scale factor (scale parameter) [92]. For the *R. pseudoacacia* classification, the subsequent OBIA was performed using class-specific features. This involved spectral information such as the mean green value, as black locust leaves have a specific light green. Furthermore, the vegetation height (digital surface model–digital terrain model) has the advantage of classifying distinctions of vegetation height for grass as well as trees in plantations and pine forests. To capture black locust trees in the shade neighbor-related pixel values as a contrast to neighboring pixels are important. The resulting classification of *R. pseudoacacia* were exported as shape files into Quantum GIS (QGIS) [93]. Furthermore, during the field survey, marked points of the last planting lines were connected. Starting from this last planting line, five zones from 0 to 10 m were generated with the QGIS software. For each zone, we intercepted and measured the *R. pseudoacacia* cover. To calculate the average distance of spreading (b), the covered area of black locust 0–10 m (A) was divided by the length of the last planting line (a), as shown in Equation (1).

$$b = \frac{A}{a} \tag{1}$$

The accuracy of classification was evaluated by selecting random samples. These randomly selected samples were then manually classified via a visual on-screen interpretation of the available image information together with additional field data. Therefore, 150–200 points were randomly distributed. The object classification was verified among *R. pseudoacacia* and non-*R. pseudoacacia*. Based on the samples, a confusion matrix was produced to evaluate the accuracy of the final classifications, including overall, user's, and producer's classification accuracies and the Kappa Index of Agreement (KIA) [94,95].

2.5.2. Classification via Deep Learning–CNN

A deep learning algorithm on single UAS images was applied to classify black locust and non-black locust. Therefore, 1000 RGB images were selected. Black locust was detected with varying light conditions (sunny and cloudy), flying altitudes (15 m, 30 m, and 100 m), UAS, and cameras (listed in Section 2.5), and thus different structures and colors. In Python [96] via Jupyter Notebook [97] the tensorflow [98] and keras [99], as well as matplotlib [100], numpy [101], pandas [102], and sklearn [103] libraries were applied for the construction and calculation of the CNN [104,105]. The images were therefore split into train (80%) and test data (20%). The applied CNN structure is presented in Figure 3. In each block, there are convolutional layers and a sub-sampling by max-pooling. In the convolutional layer in the first block, 64 filters with a size of 3 × 3 were used. In the second block 128 feature channels were integrated, and in the third block 256 feature channels were integrated. The activation function was the rectified linear unit (relu)-function [106]. As an optimization algorithm, we used "Adam" (adaptive moment estimation) [107]. Adam integrates moment and the adaptive learning rate. Furthermore, to avoid overfitting and to allow generalization, we integrated dropouts [106] after each max-pooling and fully connected-layer. The total number of trainable parameters is 4,787,330 per image. We tested our CNN architecture without and with dropout layers: (A) including 6 convolutional layers, 3 max-pooling layers, 1 flatten layer, and 2 fully connected layers; (B) including 6 convolutional layers, 3 max-pooling layers, 1 flatten layer, 2 fully connected layers, and 4 dropout layers (Figure 3). To evaluate the applied CNNs, the loss and the accuracy of the train and test data were computed.

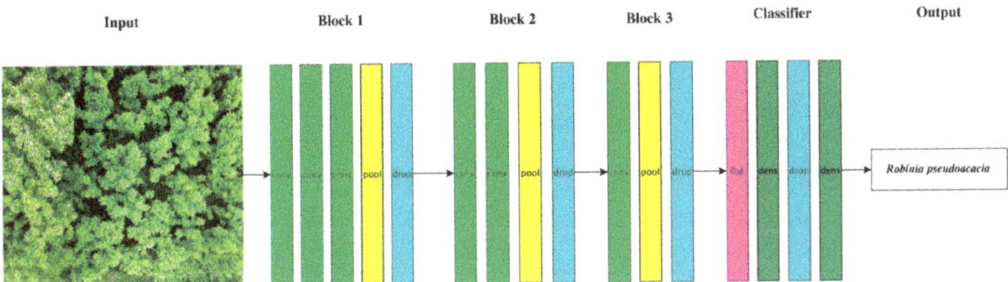

Figure 3. Convolutional neural network (CNN) structure as applied deep learning algorithm. Conv–convolutional layer (light green), pool–max-pooling layer (yellow), drop–dropout layer (blue), flat–flatten layer (pink), dens–fully connected layer (dark green).

3. Results

3.1. Seed Germination

Black locust seed germination after seeding on sandy mineral soils is presented in Figure 4. Seed germination increased with increasing scarification. Hence, seeds directly sown and watered reached 6% germination. By soaking the seeds for 24 h, 8% of the seeds sprouted. By warming the seeds for two hours at 45 °C/60 °C (113 °F/140 °F) and for two hours at −20 °C (−4 °F) as warm-cold variation, the germination reached 23%/69%. Scalding with hot water attained germination of 72% and mechanical scarification resulted in 90% of the seeds sprouting. Moreover, scalding of seeds achieved the highest standard error of the mean (Figure 4). The variance differs significantly (p-value < 0.05) by comparing all applied seed treatments. Nevertheless, the differences were non-significant (p-value > 0.05) between treatments I and II, IV and V, as well as V and VI (Table 2).

Table 2. Pairwise *t*-tests of the black locust seed germination by (I) directly sowing, (II) 24 h soaking, (III) warm-cold treatment for two hours at 45 °C and two hours at −20 °C, (IV) warm-cold treatment of two hours at 60 °C and two hours at −20 °C, (V) hot water scalding, and (VI) mechanical scarification.

Treatment	I	II	III	IV	V
II	0.3365				
III	<0.001	<0.001			
IV	<0.001	<0.001	<0.001		
V	0.0086	0.009	0.015	0.750	
VI	<0.001	<0.001	<0.001	0.011	0.274

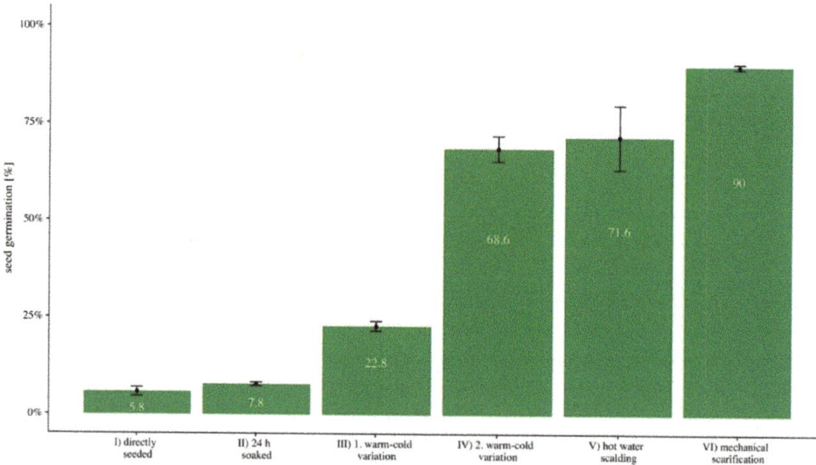

Figure 4. Black locust seed germination (%) by (I) directly sowing, (II) 24 h soaking, (III) warm-cold treatment for two hours at 45 °C and two hours at −20 °C, (IV) warm-cold treatment of two hours at 60 °C and two hours at −20 °C, (V) hot water scalding, and (VI) mechanical scarification. The error bars show the standard error of the mean per seed treatment. Database = 3000 seeds.

3.2. Stump Shoot

The average number of shoots per stump decreases with increasing age (Table 3 and Figure 5). Thereby, the average number of shoots per stump is 4.17 one year after the last harvest, shoots aged 1 year. Two years after the last harvest, an average of 3.61 shoots per stump are alive. In the third year after harvest, the value decreased to 2.18 shoots per stump. The values of the SD, SE, and CI are largest in the 3-year age class and smallest in the 2-year age class. Furthermore, there was a statistically significant difference between age classes as determined by ANOVA (Table 3). Nevertheless, the differences were non-significant between age classes 1 and 2 (p-value 0.313), but significant between age classes 1 and 3 (p-value <0.001) and between age classes 2 and 3 (p-value 0.008).

Table 3. Average number of shoots per stump and per age class as well as the plot and shoot database.

Site Abbreviation	Shoot Age	Plots	Data	Average Shoots	Shoots Per Age	SD	SE	CI	*p*-Value
WA	1	4	680	2.87	4.17	1.12	0.32	0.71	
WZ		8	1331	4.82					
BH	2	6	1000	3.98	3.61	0.61	0.16	0.35	<0.001
GU		8	1314	3.34					
BG	3	3	441	1.56	2.18	1.19	0.45	1.10	
KL		3	478	3.38					
PA		1	167	0.44					

SD: standard deviation; SE: standard error of the mean; CI: confidence interval (default 95%).

3.3. Root Suckering (Spreading) via OBIA

In Table 4 and Figure 6, the proportion of black locust classified by OBIA in the five zones is shown in percentage depending on the surrounding area: meadow, farmland, dirt road, and pine forest. Furthermore, the average distance is presented in the last row. The further away from the last planting line, the lower the proportion of black locust. Average spreading is highest to dirt road and meadow.

Table 4. Zones 0–2 m, 2–4 m, 4–6 m, 6–8 m, 8–10 m and the average distance as spreading of black locust depending on the surrounding area: meadow, farmland, dirt road, and forest.

Zone	0–2 m (%)	2–4 m (%)	4–6 m (%)	6–8 m (%)	8–10 m (%)	Average Distance (m)
Meadow	43.99	32.37	14.86	3.00	0.33	1.89
Farmland	27.45	19.60	9.78	0.91	0.07	1.16
Dirt road	41.96	37.81	26.07	6.38	0.58	2.26
Forest	40.23	25.42	7.79	1.14	0.002	1.49

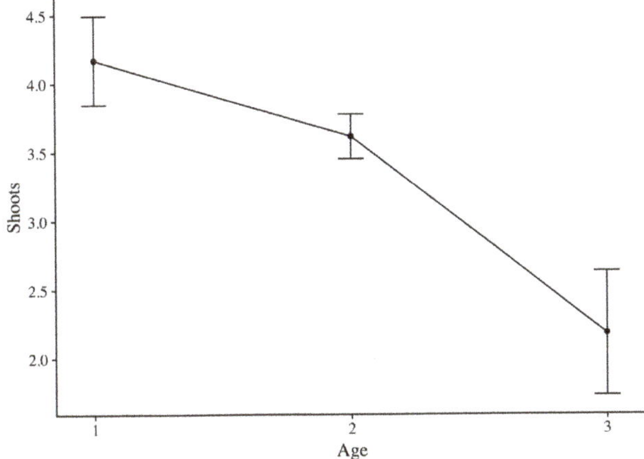

Figure 5. Average number of shoots per stump in relation to the shoot age and the standard error of the mean (data = 5244 shoots). Detailed values are presented in Table 3.

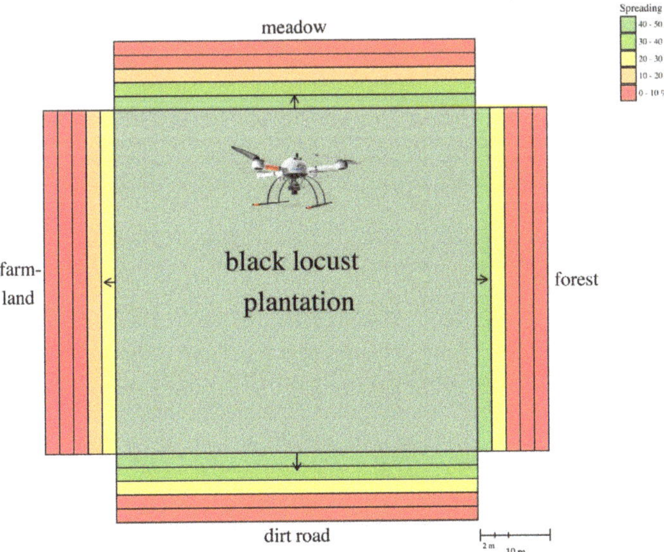

Figure 6. Spreading: outline average values of the areas; 0–2 m, 2–4 m, 4–6 m, 6–8 m, 8–10 m depending on the surrounding areas: meadow, farmland, dirt road, and pine forest. The spreading depends on the proportion of black locust; dark green stands for 40%–50%, light green indicates 30%–40%, yellow is 20%–30%, orange 10%–20% and red 0%–10%. The arrows show the average spreading distance of black locust. For an overview of the results, see Table 4. Database = 2124 m [82].

At 43.99%, the highest value in the 0–2 m zone is reached if the surrounding area is meadow, and the smallest, 27.45%, if the surrounding area is farmland. In the following four zones 2–10 m, the highest black locust quantity is measured if the surrounding area is a dirt road. Generally, the spreading in the first 2 m from the last planting line to the neighboring area is between 40%–50% if the neighboring area is meadow, dirt road, and forest. In the zone from 4–10 m, the proportion is around 0%–10% if the neighboring area is forest or farmland, and in the zone from 6–10 m, the proportion is 0%–10% if the neighboring area is meadow or dirt road. Furthermore, the highest average distance, 2.26 m, is reached if the surrounding area is a dirt road (arrow in Figure 6). At 1.89 m, the second highest average distance is measured if the surrounding area is meadow. In the case of the forest as the surrounding area, the average distance is 1.49 m. The smallest average distance, 1.16 m, is reached if the surrounding area is farmland. The overall accuracy of the OBIA analysis is 0.97 and the overall KIA is 0.93 (Table 5). The producer and user accuracy ranged between 0.96 and 0.98.

Table 5. Confusion matrix of the classification *R. pseudoacacia* and non-*R. pseudoacacia* via object-based image analysis (OBIA). KIA stands for Kappa Index of Agreement.

		Actual	
		R. pseudoacacia	*Non-R. pseudoacacia*
Predicted	*R. pseudoacacia*	177	7
	Non-R. pseudoacacia	5	174
Producer Accuracy		0.975	0.962
User Accuracy		0.963	0.972
KIA Class		0.942	0.928
Overall Accuracy		0.966	
KIA Overall		0.932	

3.4. Classification via CNN

The CNN architecture without dropout layers (A) reached a 90% test accuracy and a 99.7% train accuracy. The test loss was 0.395 and the train loss achieved 0.009. The model of the CNN (B) used for classifying *R. pseudoacacia* and Non-*R. pseudoacacia* is presented in Table 6. On the left-hand side are image examples of test data of *R. pseudoacacia* (1) classifications with varying flying altitudes, light conditions, UAS, and cameras. On the right-hand side are examples of test data of Non-*R. pseudoacacia* (0) category. The CNN's (B) accuracy (Figure 7) in classifying *R. pseudoacacia* (1) and Non-*R. pseudoacacia* (0) increased quickly. After a small number of iterations, the accuracy of the training and validation dataset increased by over 90%, ranging between 90 and 100% from iteration 5 to 50. The training accuracy reached 99.7% and the test accuracy 99.5%. The training and validation loss reduced from 0.7 to a range between 0 and 0.2. Finally, the training loss was 0.009 and the test loss 0.027. The time for fitting the model was 24.2 min, with 3.83 billion training parameters included.

Table 6. Classification examples of the test data set; *R. pseudoacacia* ((**a**), category 1) and Non-*R. pseudoacacia* ((**b**) category 0): meadow, dirt road, pine forest, and farmland. Pred is the abbreviation for Prediction. True and the green color of the labels show that the prediction (output in Figure 3) and the class (input in Figure 3) are equal.

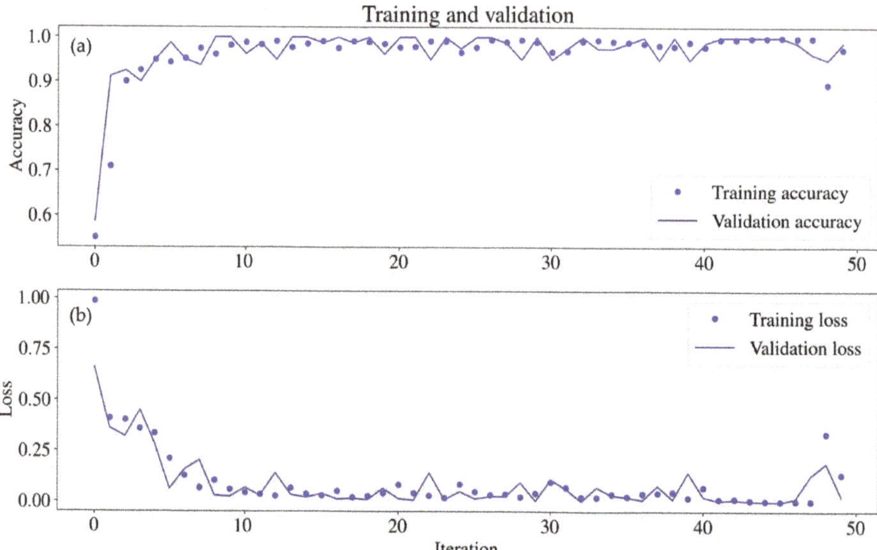

Figure 7. Accuracy (**a**) and loss (**b**) of the training and validation (=testing) data for classification of *R. pseudoacacia* and Non-*R. pseudoacacia*, depending on the iterations. Dataset = 1000 images; 80% for training and 20% for testing.

4. Discussion

Reproduction strategies of tree species are an important basis for the survival in spatially and temporally varied ecosystems. However, the reproduction and spreading of black locust in short rotation coppices in Germany have not been analyzed until now. In this study, by applying seed experiments, stump shoot analysis, and spreading measurements, we create an overview of the invasiveness potential of black locust in short rotation coppices in a temperate climate zone. Black locust reproduces generatively via seeds and vegetatively via stump shoots and root suckers. These three reproduction strategies improve the survival of black locust in many ecosystems globally [6,17,108–114]. Thereby, the reproduction and spreading increases with greater damage to the seed coat and the harvest of black locust trees. Further favorable conditions are sandy soil and high light availability.

Reproduction via seeds is a successful strategy for plants globally. Black locust seeds have a hard and impermeable seed coat [20,21]. In the present study, we observed that seed germination increased with increasing mechanical scarification. This is in line with Vines [22] and Redei et al. [23]. Moreover, the seed experiment shows that germination increases when the warm-cold variation is increased. It might be possible that when the warm-cold variation increases, micro-cracks appear, water enters, and light stimuli facilitates the growth processes. During mechanical scarification, the seed's coat is directly damaged and the seedling's development can start. Tompa and Szent-Istvany [115] conclude that hot water treatment was less effective for black locust seed germination. In our study, seeds soaked for 24 h had a 2% higher germination compared to the directly seeded category, with a 6% germination. However, there was no statistically significant difference between seed treatments I) directly seeded and II) soaked in water (18 °C/64.4 °F) for 24 h. However, scalding with hot water yielded the second highest average germination value, with germination above 70%, as well as the highest standard error of the mean. The reason could be the appearance of microcracks in some seed coats and for others the scalding stimulus might be too weak, as the seed coat permeability varied. The applicability of scalding for increasing germination is in line with Velkov [25], as well as with Bogoroditskii and Sholokhov [24]. All in all, it is important to crack and open the seed coat for successful germination.

We have to stress that seeds were collected in winter 2016 and only from two former mining areas, as the black locust trees were non-harvested, aged 8 years, and abundant seeds were collectable. It might be possible that the seed quality and germination from different former types of utilization, sites, stand ages, and years vary, as it is described by Redei et al. [17]. Therefore, depending on the seed samples, the absolute results may differ. However, the relative results might be similar. It is important to note, that flowering and reproduction via seed usually starts at the age of six years [17]. For younger short rotation coppices, the reproduction via seeds might be neglectable. Further research could focus on additional seeds selected at different sites, such as urban sites, pure and mixed forests, and on varying age, years, etc. The seed experiments could be planned as field experiments to analyze the influence of fire [31], landscape destruction [31,49], dryness, insects, diseases [116], storms, landslides [32], etc.

Building stump shoots after being damaged or harvested is a survival strategy of many broad-leafing tree species. As black locust is able to build stump shoots, grows fast, and can grow on nutrient-poor sites, this species was planted in short rotation coppices in Germany [33]. In this study, we estimate that the number of shoots per stump decreases with increasing shoot age, which differs significantly by comparing shoot age classes 1 and 3 as well as 2 and 3. This age–shoot relationship has already been observed for black locust stump shoots in forest stands [47,48]. Zeckel [47] counted 13 shoots on average, and Ertle et al. [48] observed seven shoots per stump one year after harvest, six stump shoots in the second year after harvest [47,48], and four shoots per stump in the third and fourth years after harvest [48]. Those values are higher compared with the stump shoot analysis of the present study. In short rotation coppices, we observed on average 4.2 shoots per stump one year after harvest, 3.6 shoots per stump two years after harvest, and 2.2 shoots per stump three years after harvest. One reason for the difference in the number of shoots per stump in forests compared with short rotation coppices might be the planting distance and the number of trees per hectare. Forests were planted with 2000–3000 trees per hectare, and short rotation coppices typically with 8000–10,000 trees per hectare [9]. This fact corresponds with further observations that black locust mortality increases in plantation and reclamation projects with higher density planting [117,118]. Furthermore, the range in number of shoots per stump widens as stump age and dieback of stumps increases [47,48]. Additionally, significant differences were recorded among black locust biomass production of trees (non-harvested) and stump shoots (harvested) [33] in short rotation coppices. This study is a further step in understanding the relationships of the average shoot sprout and dieback with increasing shoot age. The intensity of dieback and growth partitioning might also depend on the aboveground and belowground resources. As belowground resources, water and phosphorus were described as key drivers for competition among black locust trees and stump shoots [119]. Accordingly, by increasing water and phosphorus availability, the competition among stump shoots and the dieback of shoots increases. Additionally, the fact that the increasing shoot age correlates with a decrease in the number of shoots per stump is important for managers of short rotation coppices, as the wood biomass (quality) increases, but the shoots per stump (quantity) decreases. It could be interpreted that it might be effective to plant black locust at greater planting distances, as more shoots will survive in the first years after harvest. However, this feature could be also interpreted as a risk-minimizing and quality-improving aspect. The slenderness is reduced and the risk of breaking due to wind or weight (for example, abundant seeds) is decreased. The wood quality of the surviving shoots increases as the relative bark proportion decreases and the relative wood proportion per shoot increases. Furthermore, black locust wood differs among juvenile and adult wood [120–123]. Hence, as the black locust tree or shoot becomes older, the wood quality could improve for energy use and construction timber. Further studies might focus on the best planting distances for black locust to improve the target biomass and wood production.

As a third reproduction strategy, black locust is able to spread via root suckering. We observed that the spreading increases with increasing light availability and reduced tillage. The proportion and average spreading distance are highest if the surrounding area is a dirt road or meadow. The proportion and distance were lower if the neighboring area is a forest or farmland. The lower spreading distance to the farmland could be due to tillage by the land owner and the resulting destruction of the spreading black locust roots. The reason for lesser spreading in the neighboring area if it is a forest could be the light limitation due to the higher pine trees. As a light-demanding tree species, black locust growth is reduced if the light availability is low [124]. This is in line with previous findings discussed in Crosti et al. [50], who found that the density of black locust regeneration in Mediterranean ecosystems is strongly affected by the land cover. Accordingly, abandoned agricultural land is most prone to black locust colonization. The spread was lowest in a zone with orchards. This zone was described as very effective for controlling black locust invasion [50]. Moreover, our study shows that independent of the neighboring area, the proportion of black locust cover and spreading decreases, the longer the distance is to the plantation. Therefore, it seems that the spreading is a step-by step-process. Nevertheless, it has been found that black locust is able to spread locally up to 1 m per year [125]. Huntley [2] and Grese [126] describe that root sprouting usually begins when plants are 4 years old and increases rapidly in full sun, open areas, and particularly in sandy soil [126]. As with stump shoots, sprouting is a response to disturbance, and sprouts need sufficient light to survive [32]. Moreover, as black locust belongs to the family *Fabaceae,* it is able to fix atmospheric nitrogen via rhizobia [15,16], which results in an increasing nitrogen concentration and a change in the chemical compositions in the soil [127–129]. In Europe, particularly in Germany, nature conservation areas are often open areas, such as dry and semi-dry grasslands, which belongs to the most species-rich and endangered types of habitats [130]. To protect such areas and avoid a black locust spreading, it is important to avoid planting black locust close or next to protected areas. Long-term investigations might provide answers regarding further ecosystem modifications of black locust. Moreover, on a global scale, the spreading of atmospheric nitrogen fixing tree species is increasing [6,70,131]. This might be due to anthropogenic influences such as an increased CO_2 and nitrogen concentration in the atmosphere [132,133]. Additionally, the nitrogen concentration often increases locally due to the usage of fertilizer on farmlands [134]. The increasing spreading of atmospheric nitrogen fixing tree and shrub species that results in ecosystem-changing processes by enriching the soil with nitrogen, seems to be a manmade issue. Moreover, increasing temperature, reduced frost [2,135,136], and the dieback of tree species, such as European ash [13,14], could result in a further increase in black locust and additional non-native tree species. It is important to stress that the analyzed length to the neighboring areas varies and that mother stumps are missing in a few planting lines, which might influence the results. Therefore, to validate the results further, black locust stands with other neighboring ecosystems could be investigated.

The approach of combining UAS images with OBIA, as described in the present study, could be the foundation to create a database to appraise the spreading and ecosystem changes caused by *R. pseudoacacia* and could be further applied to non-native and native tree species. To this end, UAS allows monitoring with a higher spatial resolution of small, specific, and detailed vegetation structures [18,69,137] and avoids environmental damage. Another advantage of UAS images is the possibility to generate 3D surface models [69]. The 3D surface models in this study were important for the classification of the vegetation height of *R. pseudoacacia*, grass, and forest pine trees. The height of the black locust trees in the analyzed plantation ranged between 0.2 and 3.0 m. This height classification approach is common for tree analysis [53–55,57,58,60]. However, a limitation of current UAS images is that depending on stand density only the top of the tree or stump shoot is detectable [138]. Solutions could be the integration of high-resolution light detection and ranging (LiDAR) systems to an UAS [139] or UAS flying below the canopy as a possibility for future technology. Both could detect the deeper canopy and stand layers [139]. Along with the vegetation height, the neighbor-related pixel values and contrast were important for detecting black locust trees in shade caused by higher vegetation or cloud cover. Furthermore, the mean green values were considerable, as *R. pseudoacacia* leaves have a specific light green. The advantage of OBIA compared to pixel-based analysis is the avoidance of misclassification of single pixels [140,141]. OBIA includes geometric properties such as dimension and texture of the target tree species and creates additional options as compared to pixel-based analysis [92,142]. Nevertheless, the OBIA approach as applied in the present study relies on a labor-intensive survey, which needs expertise in image converting. This is generally time consuming and expensive [143,144]. Furthermore, mistakes due to human operation are possible [145].

In addition, machine learning algorithms are increasingly being used for data and image analysis [52,62,72–79]. The CNN applied in the current study was tested to classify single black locust images under varying conditions and attained a high test accuracy of 99.5%. However, disadvantages of CNN algorithms are that a high number of labeled training images should be available [146], as well as the difficult traceability of the used classification features [105]. Nevertheless, studies [73,145,146] have shown that when the training sample size was high, CNN tended to show better results and accuracies compared to random forest, support vector machine, and fully convolutional networks. Therefore, Liu and Abd-Elrahman [146] used 400 UAS images per object, Diegues et al. [147] applied about 700 underwater images, Abrams et al. [148] operated with 700 canopy and 800 understory images per habitat class as well as Li et al. [73] used 5,000 satellite images per category. In the present study 1000 UAS images were selected to classify black locust and non-black locust. Moreover, varying CNN architectures are existing and being further developed. We evaluate the classification accuracy of two widely used CNN architectures: (A) including 6 convolutional layers, 3 max-pooling layers, 1 flatten layer, and 2 fully connected layers as simplified VGG architecture [104]; (B) including 6 convolutional layers, 3 max-pooling layers, 1 flatten layer, 2 fully connected layers, and 4 dropout layers, which is similar to AlexNet architecture [106]. Our proposed approach (B), which includes dropout layers, achieved a higher test accuracy at 99.5% compared with the applied VGG architecture without dropout layers, which achieved a 90% test accuracy. Both CNN architectures achieved a 99.7% training accuracy. Therefore, the applied VGG architecture tend to overfit the model. Nevertheless, the applied dropout layers [106] in the applied AlexNet architecture avoided overfitting and allowed generalization with a 99.5% accuracy by modelling 1,000 images. This is in line with previous findings discussed in Li et al. [73], who found that the AlexNet architecture achieved the highest accuracy in oil palm tree detection (satellite images) compared to LeNet [149] and VGG-19 architecture. Single image classification of black locust, as applied in the present study, could be advanced in further studies by applying CNN architectures for segmentation (e.g., semantic or pedestrian segmentation) [62,145,150–152]. It is important to stress that the described CNN architectures were tested on images of black locust in a short-rotation coppice at one site in a variation of flying altitudes (15 m, 30 m, and 100 m), light conditions (sunny and cloudy), UAS (Microdrones MD4-1000 [82] and dji Mavic Pro [85]), and cameras (SONY-ILCE-5100 [87] and DJI FC220 [85]). For a comprehensive training and the applicability to different sites, further images of black locust from different age classes, health conditions, seasonal change of leaf colors (e.g., spring and autumn in Central Europe), tree heights, stand densities, competition [119], multiple mixed stands, environments, drones, etc., are needed, but this methodology offers a new possibility for faster and automatic detection of black locust and other tree species.

This study focuses on the reproduction of black locust in short rotation coppices in Germany in order to improve the estimation of the spreading and survival potential. The analysis of generative and vegetative reproduction shows that scarification increases germination, the numbers of shoots per stump decrease with time, and sprouting is influenced by light availability. Our research study gives an overview of the invasiveness potential and reproduction strategies of black locust in short rotation coppices, which is important for managing black locust effectively, for example for biomass production and nature conservation.

5. Conclusions

This study provides an overview of the reproduction strategies of black locust by analyzing the reproduction and spreading in ten short rotation coppices in Germany. The seed experiments focus on six different treatments in five iterations and show that seed germination increases with increasing warm-cold variation (23%–69%), hot water scalding (72%), and mechanical scarification (90%) of the hard and impermeable seed coat. Furthermore, after scalding with hot water, the seed germination reached the highest standard error of the mean. Moreover, the findings showed that the seed germination is less than 10% when seeds were directly seeded or soaked in water (18 °C/64.4 °F) for 24 h. Stump shoots were counted in sample plots in varying age classes (1–3 years). The numbers of shoots per stump decrease as shoot age increases, which differs significantly by comparing age classes 1 and 3, as well as 2 and 3. Spreading of root suckers was analyzed with the aid of UAS platforms and image analysis via OBIA. The spreading distance increases with increasing light availability and is decreases with tillage. Thereby, the proportion and average spreading distance are highest if the surrounding area is a dirt road or meadow. The proportion and distance were lower if the neighboring area is a forest or farmland. Furthermore, we tested a CNN model to classify black locust under varying conditions in single images (1000 images) and achieved a high accuracy of 99.5% by including 6 convolutional layers, 3 max-pooling layers, 1 flatten layer, 2 fully connected layers, and 4 dropout layers. The methodology and results presented herein provide local managers, foresters, and scientists with the opportunity to estimate reproduction and spreading of black locust and other tree species.

Author Contributions: Conceptualization, C.C., J.R.K.L., D.L. and H.P.; Investigation, C.C. and J.R.K.L.; Methodology, C.C. and J.R.K.L.; Project administration, C.C.; Software, C.C. and J.R.K.L.; Supervision, D.L. and H.P.; Validation, C.C. and J.R.K.L.; Visualization, C.C.; Writing—Original draft, C.C.; Writing—Review & editing, J.R.K.L., D.L. and H.P.

Funding: This research was supported by the University of Applied Science Erfurt (FHE), the Technical University of Munich (TUM), and the Westfälische Wilhelms University of Münster. This work was also supported by the German Research Foundation (DFG) and the Technical University of Munich (TUM) in the framework of the Open Access Publishing Program.

Acknowledgments: We would like to thank the Vattenfall-Energy Crops company, Hartmut Petrick, the Brandenburg University of Technology Cottbus-Senftenberg (BTU) for their support with analyzing the study areas. The authors would also like to thank Jan Zimmermanns for his assistance during all seed experiments and field measurements. Many thanks to Christian Rösner and Martin Rogge for their knowledge and information on black locust seeds. Moreover, we want to thank Pia Pickenbrock for her flying skills as our second pilot. We are grateful to Daniela and Cindy Carl for proofreading the article. Furthermore, many thanks to Oliver Zeigermann for his support with machine learning, especially deep learning algorithms.

Conflicts of Interest: The authors declare no conflict of interest.

Abbreviations

The following abbreviations are used in this manuscript:

Adam	Adaptive moment estimation
ANOVA	Analysis of Variance
CI	Confidence Interval
CNN	Convolutional Neural Network(s)
h	Hours
KIA	Kappa Index of Agreement
LiDAR	Light Detection and Ranging
NIR	Near-InfraRed
OBIA	Object-Based Image Analysis
QGIS	Quantum GIS
RGB	Red-Green-Blue
SD	Standard Deviation
SE	Standard Error of the Mean
SfM	Structure from Motion
UAS	Unmanned Aerial System(s)

References

1. Little, E.L. *Atlas of United States Trees*; Conifers and Important Hardwoods, US Department of Agriculture, Forest Service: Washington, DC, USA, 1971.

2. Huntley, J.C. *Robinia pseudoacacia* L. In *Silvics of North America, Vol. 2, Hardwoods*; Burns, R.M., Honkala, B.H., Eds.; USDA Foreign Agricultural Service Handbook 654: Washington, DC, USA, 1990; pp. 755–761.

3. Vadas, E. *Das Lehrrevier und der botanische Garten der königl. ung. forstl. Hochschule als Versuchsfeld*; Joerges: Selmecbánya, Slovakia, 1914; pp. 1–25.

4. Ernyey, J. Die Wanderwege der Robinie und ihre Ansiedlung in Ungarn. *Magy. Botan. Lapok* **1927**, *25*, 161–191.

5. Kolbek, J.; Vítková, M.; Vetvicka, V. Z historie stredoevropsky' ch akátin a jejich spolecenstev. From history of Central European Robinia growths and its communities. *Zpr. Ces. Bot. Spolec.* **2004**, *39*, 287–298.

6. Vítková, M.; Müllerová, J.; Sádlo, J.; Pergl, J.; Pyšek, P. Black locust (*Robinia pseudoacacia*) beloved and despised: A story of an invasive tree in Central Europe. *For. Ecol. Manag.* **2017**, *384*, 287–302. [CrossRef] [PubMed]

7. Wüstenhagen, R.; Bilharz, M. Green energy market development in Germany: Effective public policy and emerging customer demand. *Energy Policy* **2006**, *34*, 1681–1696. [CrossRef]

8. Bielefeldt, J.; Bolte, A.; Busch, G.; Dohrenbusch, A.; Kroiher, F.; Lamersdorf, N.; Schulz, U.; Stoll, B. *Energieholzproduktion in der Landwirtschaft. Chancen und Risiken aus Sicht der Natur-und Umweltschutzes*; NABU Bundesverb, 2008; pp. 17–19. Available online: https://www.nabu.de/imperia/md/content/nabude/energie/biomasse/nabu-studie_energieholz.pdf (accessed on 12 January 2017).

9. Bemmann, A.; Butler Manning, D. *Energieholzplantagen in der Landwirtschaft*; Agrimedia: Hannover, Germany, 2013; ISBN 978-3-86263-081-3.

10. Richardson, D.M.; Rejmánek, M. Trees and shrubs as invasive alien species—A global review. *Divers. Distrib.* **2011**, *17*, 788–809. [CrossRef]

11. Staska, B.; Essl, F.; Samimi, C. Density and age of invasive *Robinia pseudoacacia* modulate its impact on floodplain forests. *Basic Appl. Ecol.* **2014**, *15*, 551–558. [CrossRef]

12. Vor, T.; Bolte, A.; Spellmann, H.; Ammer, C. *Potenziale und Risiken eingeführter Baumarten—Baumartenportraits mit naturschutzfachlicher Bewertung*; Universitätsverlag Göttingen: Göttingen, Germany, 2015; pp. 277–292. ISBN 978-3-86395-240-2.

13. Willoughby, I.; Stokes, V.; Poole, J.; White, J.E.; Hodge, S.J. The potential of 44 native and non-native tree species for woodland creation on a range of contrasting sites in lowland Britain. *Forestry* **2007**, *80*, 531–553. [CrossRef]

14. Skovsgaard, J.P.; Wilhelm, G.J.; Thomsen, I.M.; Metzler, B.; Kirisits, T.; Havrdová, L.; Enderle, R.; Dobrowolska, D.; Cleary, M.; Clark, J. Silvicultural strategies for Fraxinus excelsior in response to dieback caused by *Hymenoscyphus fraxineus*. *Forestry* **2017**, *90*, 455–472. [CrossRef]

15. Hoffmann, G. Die Stickstoffbindung der Robinie (*Robinia pseudoacacia* L.). *Archiv für Forstwesen* **1961**, *10*, 627–632.

16. Hoffmann, G. Effektivität und Wirtsspezifität der Knöllchenbakterien von *Robinia pseudoacacia* L. *Archiv für Forstwesen* **1964**, *13*, 563–574.

17. Rédei, K. *Black Locust (Robinia pseudoacacia L.) Growing in Hungary*; Hungarian Forest Research Institute: Sarvar, Hungary, 2013.

18. Carl, C.; Landgraf, D.; van der Maaten-Theunissen, M.; Biber, P.; Pretzsch, H. *Robinia pseudoacacia* L. Flower Analyzed by Using Unmanned Aerial Vehicle (UAV). *Remote Sens.* **2017**, *9*, 1091. [CrossRef]

19. Marjai, Z. Az akác-magbank. *Erdészeti Lapok* **1995**, *130*, 311–313.

20. Farrar, J.L. *Trees of the Northern United States and Canada*; Blackwell Publishing: Ames, IA, USA, 1995; p. 502.

21. Schubert, J. *Lagerung und Vorbehandlung von Saatgut wichtiger Baum-und Straucharten*; LÖBF: Eberswalde-Finow, Germany, 1998.

22. Vines, R.A. *Trees, Shrubs, and Woody Vines of the Southwest*; University of Texas Press: Austin, TX, USA, 1960; Volume 1104, p. 7707.

23. Rédei, K.; Csiha, I.; Keseru, Z.; Gál, J. Influence of regeneration method on the yield and stem quality of Black locust (*Robinia pseudoacacia* L.) stands: A case study. *Acta Silv. Lign. Hung.* **2012**, *8*, 103–112. [CrossRef]

24. Bogoroditskii, I.I.; Sholokhov, L.V. German: Feuchtigkeitsregime von Robinia pseudoacacia Samen, vorbereitet für die Saat durch Vakuum-Wasser-Sättigungsmethode und durch Brühen in kochendem Wasser. *Tr. Novocherkas. Inzh.-Melior. Inta* **1975**, *16*, 115–118. (Original in Russian)

25. Velkov, D. Influence of high temperatures on the water regime and viability of black locust (*Robinia pseudoacacia* L.) seeds. In Proceedings of the International Symposium on Seed Physiology of Woody Plants at Kornik, Panstwowe, Wydawnictwo Naukowe, Warszawa-Poznan, Poland, 3–8 September 1968; pp. 111–119.

26. Hull, J.C.; Scott, R.C. Plant Succession on Debris Avalanches of Nelson County, Virginia. *Castanea* **1982**, *47*, 158–176.

27. Martin, W.H. *The Role and History of Fire in the Daniel Boone National Forest*; Final Report; U.S. Department of Agriculture, Forest Service, Daniel Boone National Forest: Winchester, KY, USA, 1990; p. 131.

28. Keresztesi, B. *The Black Locust*; Akadémiai Kiadó: Budapest, Hungary, 1988.

29. Voss, E.G. *Michigan Flora. Part II. Dicots (Saururaceae–Cornaceae)*; Bull. 59; Cranbrook Institute of Science: Bloomfield Hills, MI, USA; University of Michigan Herbarium: Ann Arbor, MI, USA, 1985; p. 724.

30. Harrod, J.C.; Harmon, M.E.; White, P.S. Post-fire succession and 20th century reduction in fire frequency on xeric southern Appalachian sites. *J. Veg. Sci.* **2000**, *11*, 465–472. [CrossRef]

31. Elliott, K.J.; Vose, J.M.; Clinton, B.D.; Knoepp, J.D. Effects of understory burning in a mesic mixed-oak forest of the southern Appalachians. In *Fire in Temperate, Boreal, and Montane Ecosystems: Proceedings of the 22nd Tall Timbers Fire Ecology Conference: An International Symposium, Kananaskis Village, AB, Canada, 15–18 October 2001*; Tall Timbers Research: Tallahassee, FL, USA, 2004; pp. 272–283.

32. Stone, K.R. Robinia pseudoacacia. In *Fire Effects Information System*; U.S. Department of Agriculture, Forest Service, Rocky Mountain Research Station, Fire Sciences Laboratory (Producer), 2009; Available online: https://www.fs.fed.us/database/feis/plants/tree/robpse/all.html (accessed on 11 December 2018).

33. Carl, C.; Biber, P.; Landgraf, D.; Buras, A.; Pretzsch, H. Allometric Models to Predict Aboveground Woody Biomass of Black Locust (*Robinia pseudoacacia* L.) in Short Rotation Coppice in Previous Mining and Agricultural Areas in Germany. *Forests* **2017**, *8*, 328. [CrossRef]

34. Bernetti, G. Macchia coppices with prevalent Quercus ilex in Tuscany. In *L'auxometria dei boschi cedui Italiani*; Forestale e Montane: Tuscany, Italy, 1990; Volume 35, pp. 1–24.

35. Reich, P.B.; Teskey, R.O.; Johnson, P.S.; Hinckley, T.M. Periodic root and shoot growth in oak. *For. Sci.* **1980**, *26*, 590–598. [CrossRef]

36. Cobb, S.W.; Miller, A.E.; Zahner, R. Recurrent shoot flushes in scarlet oak stump sprouts. *For. Sci.* **1985**, *31*, 725–730. [CrossRef]

37. Dimitrov, E.P.; Stiptsov, V. Yield table for coppice stands of *Quercus cerris* in Bulgaria. *Gorsko Stopanstvo* **1991**, *47*, 13–14.

38. Beky, A. Yield of sessile oak coppice stands (*Quercus petraea*). *Erdeszeti-Kutatasok* **1991**, *82–83*, 176–192.

39. Tatoni, T.; Roche, P. Comparison of old-field and forest revegetation dynamics in Provence. *J. Veg. Sci.* **1994**, *5*, 295–302. [CrossRef]

40. Cañellas, I.; Montero, G.; Bachiller, A. Transformation of quejigo oak (*Quercus faginea* Lam.) coppice forest into high forest by thinning. *Ann. Ist. Sper. Selvic.* **1996**, *27*, 143–147.

41. Cinnirella, S.; Iovino, F.; Porto, P.; Ferro, V. Anti-erosive effectiveness of Eucalyptus coppices through the cover management factor estimate. *Hydrol. Process* **1998**, *12*, 635–649. [CrossRef]

42. Chatziphilippidis, G.; Spyroglou, G. Modelling the Growth of Quercus frainetto in Greece. In *Sustainable Forest Management–Growth Models for Europe*; Springer: Berlin/Heidelberg, Germany, 2006; pp. 373–393.

43. Fonti, P.; Cherubini, P.; Rigling, A.; Weber, P.; Biging, G. Tree rings show competition dynamics in abandoned Castanea sativa coppices after land-use changes. *J. Veg. Sci.* **2006**, *17*, 103–112. [CrossRef]

44. Salazar-García, S.; Cossio-Vargas, L.E.; Lovatt, C.J.; González-Durán, I.J.; Pérez-Barraza, M.H. Crop Load Affects Vegetative Growth Flushes and Shoot Age Influences Irreversible Commitment to Flowering of 'Hass' Avocado. *HortScience* **2006**, *41*, 1541–1546.

45. Kneifl, M.; Kadavý, J.; Knott, R. Gross value yield potential of coppice, high forest and model conversion of high forest to coppice on best sites. *J. For. Sci.* **2011**, *57*, 536–546. [CrossRef]

46. Razakamanarivo, R.H.; Razakavololona, A.; Razafindrakoto, M.A.; Vieilledent, G.; Albrecht, A. Below-ground biomass production and allometric relationships of eucalyptus coppice plantation in the central highlands of Madagascar. *Biomass Bioenergy* **2012**, *45*, 1–10. [CrossRef]

47. Zeckel, C. Betrachtung des Ertragspotenzials von Stockausschlägen der Robinie (*Robinia pseudoacacia* L.) von verschiedenen Waldstandorten geogenen und anthorpogenen Ausgangssubstrates in der Niederlausitz unter Berücksichtigung ihrer energetischen Nutzung. Diploma Thesis, Brandenburg University of Technology, Senftenberg, Germany, 2007; pp. 38–39.

48. Ertle, C.; Böcker, L.; Landgraf, D. Wuchspotenzial von Stockausschlägen der Robinie. *AFZ-Der Wald* **2008**, *63*, 994–995.

49. Dooley, T. Lessons learned from eleven years of prescribed fire at the Albany Pine Bush Preserve. In *Using Fire to Control Invasive Plants: What's New, What Works in the Northeast?—2003 Workshop Proceedings*; University of New Hampshire, Cooperative Extension: Portsmouth, NH, USA; Durham, NH, USA, 2003; pp. 7–10. Available online: http://extension.unh.edu/resources/files/Resource000412_Rep434.pdf (accessed on 10 June 2016).

50. Crosti, R.; Agrillo, E.; Ciccarese, L.; Guarino, R.; Paris, P.; Testi, A. Assessing escapes from short rotation plantations of the invasive tree species *Robinia pseudoacacia* L. in Mediterranean ecosystems: A study in central Italy. *IFOREST* **2016**, e1–e8. [CrossRef]

51. Ustin, S.L.; DiPietro, D.; Olmstead, K.; Underwood, E.; Scheer, G.J. Hyperspectral remote sensing for invasive species detection and mapping. In Proceedings of the 2002 IEEE International Geoscience and Remote Sensing Symposium, IGARSS'02, Toronto, ON, Canada, 24–28 June 2002; Volume 3, pp. 1658–1660.

52. Zhao, Q.; Wang, F.; Zhao, J.; Zhou, J.; Yu, S.; Zhao, Z. Estimating Forest Canopy Cover in Black Locust (*Robinia pseudoacacia* L.) Plantations on the Loess Plateau Using Random Forest. *Forests* **2018**, *9*, 623. [CrossRef]

53. Wallace, L.; Lucieer, A.; Malenovský, Z.; Turner, D.; Vopěnka, P. Assessment of forest structure using two UAV techniques: A comparison of airborne laser scanning and structure from motion (SfM) point clouds. *Forests* **2016**, *7*, 62. [CrossRef]

54. Ota, T.; Ogawa, M.; Mizoue, N.; Fukumoto, K.; Yoshida, S. Forest Structure Estimation from a UAV-Based Photogrammetric Point Cloud in Managed Temperate Coniferous Forests. *Forests* **2017**, *8*, 343. [CrossRef]

55. Mlambo, R.; Woodhouse, I.H.; Gerard, F.; Anderson, K. Structure from Motion (SfM) photogrammetry with drone data: A low cost method for monitoring greenhouse gas emissions from forests in developing countries. *Forests* **2017**, *8*, 68. [CrossRef]

56. Mohan, M.; Silva, C.A.; Klauberg, C.; Jat, P.; Catts, G.; Cardil, A.; Hudak, A.T.; Dia, M. Individual tree detection from unmanned aerial vehicle (UAV) derived canopy height model in an open canopy mixed conifer forest. *Forests* **2017**, *8*, 340. [CrossRef]

57. Guerra-Hernández, J.; González-Ferreiro, E.; Monleón, V.J.; Faias, S.P.; Tomé, M.; Díaz-Varela, R.A. Use of Multi-Temporal UAV-Derived Imagery for Estimating Individual Tree Growth in Pinus pinea Stands. *Forests* **2017**, *8*, 300. [CrossRef]

58. Dempewolf, J.; Nagol, J.; Hein, S.; Thiel, C.; Zimmermann, R. Measurement of within-season tree height growth in a mixed forest stand using UAV imagery. *Forests* **2017**, *8*, 231. [CrossRef]

59. Qiu, Z.; Feng, Z.K.; Wang, M.; Li, Z.; Lu, C. Application of UAV Photogrammetric System for Monitoring Ancient Tree Communities in Beijing. *Forests* **2018**, *9*, 735. [CrossRef]

60. Alonzo, M.; Andersen, H.E.; Morton, D.C.; Cook, B.D. Quantifying Boreal Forest Structure and Composition Using UAV Structure from Motion. *Forests* **2018**, *9*, 119. [CrossRef]

61. Feduck, C.; McDermid, G.; Castilla, G. Detection of coniferous seedlings in UAV imagery. *Forests* **2018**, *9*, 432. [CrossRef]

62. Morales, G.; Kemper, G.; Sevillano, G.; Arteaga, D.; Ortega, I.; Telles, J. Automatic Segmentation of Mauritia flexuosa in Unmanned Aerial Vehicle (UAV) Imagery Using Deep Learning. *Forests* **2018**, *9*, 736. [CrossRef]

63. Fraser, B.; Congalton, R. Evaluating the Effectiveness of Unmanned Aerial Systems (UAS) for Collecting Thematic Map Accuracy Assessment Reference Data in New England Forests. *Forests* **2019**, *10*, 24. [CrossRef]

64. Knipling, E.B. Physical and physiological basis for the reflectance of visible and near-infrared radiation from vegetation. *Remote Sens. Environ.* **1970**, *1*, 155–159. [CrossRef]

65. Tucker, C.J. Red and photographic infrared linear combinations for monitoring vegetation. *Remote Sens. Environ.* **1979**, *8*, 127–150. [CrossRef]

66. Qi, J.; Chehbouni, A.; Huete, A.R.; Kerr, Y.H.; Sorooshian, S. A modified soil adjusted vegetation index. *Remote Sens. Environ.* **1994**, *48*, 119–126. [CrossRef]

67. Rogan, J.; Franklin, J.; Roberts, D.A. A comparison of methods for monitoring multitemporal vegetation change using Thematic Mapper imagery. *Remote Sens. Environ.* **2002**, *80*, 143–156. [CrossRef]

68. Gitelson, A.A.; Kaufman, Y.J.; Stark, R.; Rundquist, D. Novel algorithms for remote estimation of vegetation fraction. *Remote Sens. Environ.* **2002**, *80*, 76–87. [CrossRef]

69. Tang, L.; Shao, G. Drone remote sensing for forestry research and practices. *J. For. Res.* **2015**, *26*, 791–797. [CrossRef]

70. Lehmann, J.R.; Prinz, T.; Ziller, S.R.; Thiele, J.; Heringer, G.; Meira-Neto, J.A.; Buttschardt, T.K. Open-source processing and analysis of aerial imagery acquired with a low-cost unmanned aerial system to support invasive plant management. *Front. Environ. Sci.* **2017**, *5*, 44. [CrossRef]

71. Müllerová, J.; Bartaloš, T.; Brůna, J.; Dvořák, P.; Vítková, M. Unmanned aircraft in nature conservation: An example from plant invasions. *Int. J. Remote Sens.* **2017**, *38*, 2177–2198. [CrossRef]

72. Daliakopoulos, I.N.; Katsanevakis, S.; Moustakas, A. Spatial downscaling of alien species presences using machine learning. *Front. Earth Sci.* **2017**, *5*, 60. [CrossRef]

73. Li, W.; Dong, R.; Fu, H. Large-Scale Oil Palm Tree Detection from High-Resolution Satellite Images Using Two-Stage Convolutional Neural Networks. *Remote Sens.* **2019**, *11*, 11. [CrossRef]

74. Denize, J.; Hubert-Moy, L.; Betbeder, J.; Corgne, S.; Baudry, J.; Pottier, E. Evaluation of Using Sentinel-1 and-2 Time-Series to Identify Winter Land Use in Agricultural Landscapes. *Remote Sens.* **2019**, *11*, 37. [CrossRef]

75. Zhou, K.; Lindenbergh, R.; Gorte, B. Automatic Shadow Detection in Urban Very-High-Resolution Images Using Existing 3D Models for Free Training. *Remote Sens.* **2019**, *11*, 72. [CrossRef]

76. Wei, S.; Zhang, H.; Wang, C.; Wang, Y.; Xu, L. Multi-Temporal SAR Data Large-Scale Crop Mapping Based on U-Net Model. *Remote Sens.* **2019**, *11*, 68. [CrossRef]

77. Duarte-Carvajalino, J.; Alzate, D.; Ramirez, A.; Santa-Sepulveda, J.; Fajardo-Rojas, A.; Soto-Suárez, M. Evaluating Late Blight Severity in Potato Crops Using Unmanned Aerial Vehicles and Machine Learning Algorithms. *Remote Sens.* **2018**, *10*, 1513. [CrossRef]

78. Duarte, D.; Nex, F.; Kerle, N.; Vosselman, G. Multi-Resolution Feature Fusion for Image Classification of Building Damages with Convolutional Neural Networks. *Remote Sens.* **2018**, *10*, 1636. [CrossRef]

79. Lu, T.; Ming, D.; Lin, X.; Hong, Z.; Bai, X.; Fang, J. Detecting building edges from high spatial resolution remote sensing imagery using richer convolution features network. *Remote Sens.* **2018**, *10*, 1496. [CrossRef]

80. DWD. Deutscher Wetterdienst Archiv Monats- und Tageswerte. 2018. Available online: http://www.dwd.de (accessed on 10 June 2018).

81. ESRI. ArcGIS 9.2. © Environmental Systems Research Institute. 2007. Available online: http://www.esri.com/software/arcgis/eval-help/arcgis-92 (accessed on 19 March 2017).

82. Microdrones. MD4-1000. 2018. Available online: https://www.microdrones.com/de/integrated-systems/mdmapper1000dg/ (accessed on 1 October 2018).

83. R Core Team. R: A Language and Environment for Statistical Computing—Version R 3.3.2 GUI 1.68. Available online: https://www.R-project.org/ (accessed on 2 November 2016).

84. Hain, J. *Statistik mit R: Grundlagen der Datenanalyse*; RRZN-Handbook, Regionales Rechenzentrum für Niedersachsen/Leipnitz Universität Hannover/Lehrstuhl für Mathematik VIII (Statistik) der Universität Würzburg: Würzburg/Hannover, Germany, 2011; pp. 176–187.

85. SZ DJI Technology Co., Ltd. DJI MAVIC Pro. 2018. Available online: https://www.dji.com/de/mavic (accessed on 10 October 2018).

86. Mapir Camera. Mapir Survey 3 Camera. 2018. Available online: https://www.mapir.camera/collections/survey3 (accessed on 12 June 2018).

87. Sony Europe Limited. SONY-ILCE-5100. 2018. Available online: https://www.sony.de/electronics/wechselobjektivkameras/ilce-5100-body-kit (accessed on 12 August 2018).

88. Trimble. DGPS—Trimble. 2018. Available online: https://www.trimble.com/gps_tutorial/dgps.aspx (accessed on 10 June 2018).

89. Dandois, J.P.; Olano, M.; Ellis, E.C. Optimal altitude, overlap, and weather conditions for computer vision UAV estimates of forest structure. *Remote Sens.* **2015**, *7*, 13895–13920. [CrossRef]

90. Pix4D S.A. Pix4D Mapper. 2018. Available online: https://cloud.pix4d.com/store/?=&solution=pro#solution_pro (accessed on 1 October 2018).

91. Trimble. eCognition Developer Software. 2018. Available online: http://www.ecognition.com/suite/ecognition-developer (accessed on 1 May 2018).

92. Blaschke, T. Object based image analysis for remote sensing. *ISPRS* **2010**, *65*, 2–16. [CrossRef]

93. QGIS Development Team. QGIS Geographic Information System. *Open Source Geospatial Foundation 2018*. Available online: http://qgis.osgeo.org (accessed on 8 May 2018).

94. Congalton, R.G.; Green, K. *Assessing the Accuracy of Remotely Sensed Data—Principles and Practices*, 2nd ed.; CRC Press, Taylor & Francis Group: Boca Raton, FL, USA, 2009.

95. Landis, J.R.; Koch, G.G. The measurement of observer agreement for categorical data. *Biometrics* **1977**, *33*, 159–174. [CrossRef] [PubMed]

96. Python Software Foundation. Python Language Reference, Version 3.7.1. 2018. Available online: http://www.python.org (accessed on 10 August 2018).

97. Kluyver, T.; Ragan-Kelley, B.; Pérez, F.; Granger, B.E.; Bussonnier, M.; Frederic, J.; Kelly, K.; Hamrick, J.; Grout, J.; Corlay, S.; et al. Jupyter Development Team. Jupyter Notebooks—A publishing format for reproducible computational workflows. *ELPUB* **2016**, 87–90. [CrossRef]

98. Abadi, M.; Agarwal, A.; Barham, P.; Brevdo, E.; Chen, Z.; Citro, C.; Corrado, G.S.; Davis, A.; Dean, J.; Devin, M.; et al. TensorFlow: Large-scale Machine Learning on Heterogeneous Systems. 2015. Available online: Tensorflow.org (accessed on 10 August 2018).

99. Chollet, F. Keras: The Python Deep Learning Library. 2015. Available online: https://keras.io (accessed on 10 August 2018).

100. Hunter, J.D. Matplotlib: A 2D graphics environment. *Comput. Sci. Eng.* **2007**, *9*, 90–95. [CrossRef]

101. Oliphant, T.E. *A Guide to NumPy*; Trelgol Publishing: Spanish Fork, UT, USA, 2006; Volume 1, p. 85.

102. McKinney, W. Data structures for statistical computing in python. In Proceedings of the 9th Python in Science Conference, Austin, TX, USA, 28 June–3 July 2010; Volume 445, pp. 51–56.

103. Pedregosa, F.; Varoquaux, G.; Gramfort, A.; Michel, V.; Thirion, B.; Grisel, O.; Blondel, M.; Prettenhofer, P.; Weiss, R.; Dubourg, V.; et al. Scikit-learn: Machine Learning in Python. *J. Mach. Learn. Res.* **2011**, *12*, 2825–2830.

104. Simonyan, K.; Zisserman, A. Very deep convolutional networks for large-scale image recognition. *arXiv*, 2014; arXiv:1409.1556.

105. Nguyen, C.N.; Zeigermann, O. *Machine Learning, kurz und gut*; O'Reillys Taschenbibliothek: Heidelberg, Germany, 2018; ISBN 978-3-96009-052-6.

106. Krizhevsky, A.; Sutskever, I.; Hinton, G.E. Imagenet classification with deep convolutional neural networks. In Proceedings of the Advances in Neural Information Processing Systems, Lake Tahoe, NV, USA, 3–6 December 2012; pp. 1097–1105.

107. Kingma, D.P.; Ba, J. Adam: A method for stochastic optimization. *arXiv*, 2014; arXiv:1412.6980.

108. Makkar, H.P.S.; Becker, K. Do tannins in leaves of trees and shrubs from African and Himalayan regions differ in level and activity? *Agrofor. Syst.* **1998**, *40*, 59–68. [CrossRef]

109. Auld, B.; Morita, H.; Nishida, T.; Ito, M.; Michael, P. Shared exotica: Plant invasions of Japan and south eastern Australia. *Cunninghamia* **2003**, *8*, 147–152.

110. Fuentes, N.; Ugarte, E.; Kühn, I.; Klotz, S. Alien plants in southern South America. A framework for evaluation and management of mutual risk of invasion between Chile and Argentina. *Biol. Invas.* **2010**, *12*, 3227–3236. [CrossRef]

111. Cierjacks, A.; Kowarik, I.; Joshi, J.; Hempel, S.; Ristow, M.; Lippe, M.; Weber, E. Biological flora of the British Isles: *Robinia pseudoacacia*. *J. Ecol.* **2013**, *101*, 1623–1640. [CrossRef]

112. Li, T.; Liu, G. Age-related changes of carbon accumulation and allocation in plants and soil of black locust forest on Loess Plateau in Ansai County, Shaanxi Province of China. *Chin. Geogr. Sci.* **2014**, *24*, 414–422. [CrossRef]

113. Akatov, V.V.; Akatova, T.V.; Shadzhe, A.E. *Robinia pseudoacacia* L. in the Western Caucasus. *Russ. J. Biol. Invas.* **2016**, *7*, 105–118. [CrossRef]

114. Lei, J.; Nan, L.; Bojie, F.; Guangyao, G.; Shuai, W.; Tiantian, J.; Liwei, Z.; Jianbo, L.; Di, Z. Comparison of transpiration between different aged black locust (*Robinia pseudoacacia*) trees on the semi-arid Loess Plateau, China. *J. Arid. Land.* **2016**, *8*, 604–617. [CrossRef]

115. Tompa, K.; Szent-Istvany, A. *German: Die Vorbereitung der Robiniensamen zur Saat mit Hilfe des Scheiben-Skarifikators*; Erdömernöki Föiskola, Erdötelepites-Es Fasitastani Tanszek, Sopron, Sonderdr. O. J.: Sopron, Hungary, 1963; pp. 105–121.

116. Czarapata, E.J. *Invasive Plants of the Upper Midwest: An Illustrated Guide to Their Identification and Control*; The University of Wisconsin Press: Madison, WI, USA, 2005; p. 215.

117. Clark, F.B. *Forest Planting on Strip-Mined Land*; Technical Paper No. 141; U.S. Department of Agriculture, Forest Service, Central States Forest Experiment Station: Columbus, OH, USA, 1954; p. 33.

118. Geyer, W.A. Biomass yield potential of short-rotation hardwoods in the Great Plains. *Biomass* **1989**, *20*, 167–175. [CrossRef]

119. Carl, C.; Biber, P.; Veste, M.; Landgraf, D.; Pretzsch, H. Key drivers of competition and growth partitioning among *Robinia pseudoacacia* L. trees. *For. Ecol. Manag.* **2018**, *430*, 86–93. [CrossRef]

120. Dünisch, O.; Koch, G.; Dreiner, K. Verunsicherung uber die Eigenschaften von Robinienholz. *Holz-Zentralblatt* **2007**, *39*, 1061–1062.

121. Koch, G.; Dünisch, O. *Juvenile wood in Robinie—Qualität von Robinienholz (Ro-binia pseudoacacia L.) und Folgerungen für Holzbearbeitung und Produktqualität*; Abschlussbericht für das DGfH/AIF-Forschungsvorhaben, Fraunhofer IRB Verl.: Stuttgart, Germany, 2008.

122. Dünisch, O.; Richter, H.-G.; Koch, G. Wood properties of juvenile and mature heartwood in *Robinia pseudoacacia* L. *Wood Sci. Technol.* **2010**, *44*, 301–313. [CrossRef]

123. Latorraca, J.V.F.; Dünisch, O.; Koch, G. Chemical composition and natural durability of juvenile and mature heartwood of *Robinia pseudoacacia* L. *Anais da Academia Brasileira de Ciências* **2011**, *83*, 1059–1068. [CrossRef] [PubMed]

124. Xu, F.; Guo, W.; Wang, R.; Xu, W.; Du, N.; Wang, Y. Leaf movement and photosynthetic plasticity of black locust (*Robinia pseudoacacia*) alleviate stress under different light and water conditions. *Acta Physiol. Plant* **2009**, *31*, 553–563. [CrossRef]

125. Kowarik, I. Funktionen klonalen Wachstums von Bäumen bei der Brachflächen-Sukzession unter besonderer Beachtung von *Robinia pseudoacacia*. *Verh. Ges. Ökologie* **1996**, *26*, 173–181.

126. Grese, R. The landscape architect and problem exotic plants. In Proceedings of the American Society of Landscape Architects' Open Committee on Reclamation: Reclamation Diversity, San Diego, CA, USA, 29 October 1991; Burley, J.B., Ed.; American Society of Landscape Architects: Washington, DC, USA, 1991; pp. 7–15.

127. Van Miegroet, H.; Cole, D.W. The impact of nitrification on soil acidification and cation leaching in a red alder ecosystem. *J. Environ. Qual.* **1984**, *13*, 586–590. [CrossRef]

128. Montagnini, F.; Haines, B.; Swank, W.T. Soil-solution chemistry in black locust, pine mixed-hardwoods and oak hickory forest stands in the Southern Appalachians, USA. *For. Ecol. Manag.* **1991**, *40*, 199–208. [CrossRef]

129. Malcolm, G.M.; Bush, D.S.; Rice, S.K. Soil nitrogen conditions approach preinvasion levels following restoration of nitrogen-fixing black locust (*Robinia pseudoacacia*) stands in a pine-oak ecosystem. *Restor. Ecol.* **2008**, *16*, 70–78. [CrossRef]

130. Fischer, M.; Stöcklin, J. Local extinctions of plants in remnants of extensively used calcareous grasslands 1950–1985. *Conserv. Biol.* **1997**, *11*, 727–737. [CrossRef]

131. de Sá, N.C.; Castro, P.; Carvalho, S.; Marchante, E.; López-Núñez, F.A.; Marchante, H. Mapping the Flowering of an Invasive Plant Using Unmanned Aerial Vehicles: Is There Potential for Biocontrol Monitoring? *Front. Plant Sci.* **2018**, *9*, 293. [CrossRef] [PubMed]

132. Pachauri, R.K.; Allen, M.R.; Barros, V.R.; Broome, J.; Cramer, W.; Christ, R.; Dubash, N.K. *Climate Change 2014: Synthesis Report. Contribution of Working Groups I, II and III to the Fifth Assessment Report of the Intergovernmental Panel on Climate Change*; IPCC: Geneva, Switzerland, 2014.

133. Oblack, R. Nitrogen—Gases in the Atmosphere. *ThoughtCo.* 22 June 2018. Available online: Thoughtco.com/nitrogen-in-the-atmosphere-3444094 (accessed on 9 August 2018).

134. Gordon, N.; Holland, E. Nitrogen in the Earth System—Background on the Science, People, and Issues Involved in Nitrogen Cycle Research. 2015. Available online: https://www2.ucar.edu/news/backgrounders/nitrogen-earth-system (accessed on 9 August 2018).

135. Strode, D.D. Black locust/*Robinia pseudoacacia* L. In *Woody Plants as Wildlife Food Species*. SO-16; U.S. Department of Agriculture, Forest Service, Southern Forest Experiment Station: Atlanta, GA, USA, 1977; pp. 215–216.

136. Bartha, D.; Csiszár, Á.; Zsigmond, V. Black locust (*Robinia pseudoacacia* L.). In *The Most Invasive Plants in Hungary*; Botta-Dukát, Z., Balogh, L., Eds.; Institute of Ecology and Botany, Hungarian Academy of Sciences: Vácrátót, Hungary, 2008; pp. 63–76.

137. Anderson, K.; Gaston, K.J. Lightweight unmanned aerial vehicles will revolutionize spatial ecology. *Front. Ecol. Environ.* **2013**, *11*, 138–146. [CrossRef]

138. Nevalainen, O.; Honkavaara, E.; Tuominen, S.; Viljanen, N.; Hakala, T.; Yu, X.; Hyyppä, J.; Heikki, S.; Pölönen, I.; Imai, N.N.; et al. Individual tree detection and classification with UAV-based photogrammetric point clouds and hyperspectral imaging. *Remote Sens.* **2017**, *9*, 185. [CrossRef]

139. Wallace, L.; Lucieer, A.; Watson, C.; Turner, D. Development of a UAV-LiDAR system with application to forest inventory. *Remote Sens.* **2012**, *4*, 1519–1543. [CrossRef]

140. Yan, G.; Mas, J.F.; Maathuis, B.H.P.; Xiangmin, Z.; Van Dijk, P.M. Comparison of pixel-based and object-oriented image classification approaches—A case study in a coal fire area, Wuda, Inner Mongolia, China. *Int. J. Remote Sens.* **2006**, *27*, 4039–4055. [CrossRef]

141. Weih, R.C.; Riggan, N.D. Object-based classification vs. pixel-based classification: Comparative importance of multi-resolution imagery. The International Archives of the Photogrammetry. *Remote Sens. Spat. Inf. Sci.* **2010**, *38*, C7.

142. Lehmann, J.R.; Münchberger, W.; Knoth, C.; Blodau, C.; Nieberding, F.; Prinz, T.; Pancotto, V.A.; Kleinebecker, T. High-resolution Classification of South Patagonian Peat Bog Microforms Reveals Potential Gaps in Up-scaled CH4 Fluxes by use of Unmanned Aerial System (UAS) and CIR Imagery. *Remote Sens.* **2016**, *8*, 173. [CrossRef]

143. Ammour, N.; Alhichri, H.; Bazi, Y.; Benjdira, B.; Alajlan, N.; Zuair, M. Deep learning approach for car detection in UAV imagery. *Remote Sens.* **2017**, *9*, 312. [CrossRef]

144. Cao, R.; Zhu, J.; Tu, W.; Li, Q.; Cao, J.; Liu, B.; Zhang, Q.; Qiu, G. Integrating Aerial and Street View Images for Urban Land Use Classification. *Remote Sens.* **2018**, *10*, 1553. [CrossRef]

145. Xu, Y.; Xie, Z.; Feng, Y.; Chen, Z. Road Extraction from High-Resolution Remote Sensing Imagery Using Deep Learning. *Remote Sens.* **2018**, *10*, 1461. [CrossRef]

146. Liu, T.; Abd-Elrahman, A. An Object-Based Image Analysis Method for Enhancing Classification of Land Covers Using Fully Convolutional Networks and Multi-View Images of Small Unmanned Aerial System. *Remote Sens.* **2018**, *10*, 457. [CrossRef]

147. Diegues, A.; Pinto, J.; Ribeiro, P. Automatic Habitat Mapping using Convolutional Neural Networks. *IEEE OES AUV* **2018**. Available online: https://www.researchgate.net/profile/Jose_Pinto17/publication/330449471_Automatic_Habitat_Mapping_using_Convolutional_Neural_Networks/links/5c408fc9458515a4c72d24b7/Automatic-Habitat-Mapping-using-Convolutional-Neural-Networks.pdf (accessed on 12 August 2018).

148. Abrams, J.F.; Vashishtha, A.; Wong, S.T.; Nguyen, A.; Mohamed, A.; Wieser, S.; Kuijper, A.; Wilting, A.; Mukhopadhyay, A. Habitat-Net: Segmentation of habitat images using deep learning. *bioRxiv* **2018**, 483222. [CrossRef]

149. LeCun, Y.; Bottou, L.; Bengio, Y.; Haffner, P. Gradient-based learning applied to document recognition. *Proc. IEEE* **1998**, *86*, 2278–2324. [CrossRef]

150. Paszke, A.; Chaurasia, A.; Kim, S.; Culurciello, E. Enet: A deep neural network architecture for real-time semantic segmentation. *arXiv*, 2016; arXiv:1606.02147.

151. Garcia-Garcia, A.; Orts-Escolano, S.; Oprea, S.; Villena-Martinez, V.; Garcia-Rodriguez, J. A review on deep learning techniques applied to semantic segmentation. *arXiv*, 2017; arXiv:1704.06857.

152. Flohr, F.; Gavrila, D. PedCut: An iterative framework for pedestrian segmentation combining shape models and multiple data cues. *BMVC* **2013**. [CrossRef]

 *forests*

Article

Evaluating the Effectiveness of Unmanned Aerial Systems (UAS) for Collecting Thematic Map Accuracy Assessment Reference Data in New England Forests

Benjamin T. Fraser * and Russell G. Congalton

Department of Natural Resources and the Environment, University of New Hampshire, 56 College Road, Durham, NH 03824, USA; Russ.Congalton@unh.edu

* Correspondence: btc28@wildcats.unh.edu

Received: 27 November 2018; Accepted: 27 December 2018; Published: 3 January 2019

Abstract: Thematic mapping provides today's analysts with an essential geospatial science tool for conveying spatial information. The advancement of remote sensing and computer science technologies has provided classification methods for mapping at both pixel-based and object-based analysis, for increasingly complex environments. These thematic maps then serve as vital resources for a variety of research and management needs. However, to properly use the resulting thematic map as a decision-making support tool, an assessment of map accuracy must be performed. The methods for assessing thematic accuracy have coalesced into a site-specific multivariate analysis of error, measuring uncertainty in relation to an established reality known as reference data. Ensuring statistical validity, access and time constraints, and immense costs limit the collection of reference data in many projects. Therefore, this research proposes evaluating the feasibility of adopting the low-cost, flexible, high-resolution sensor-capable Unmanned Aerial Systems (UAS, UAV, or Drone) platform for collecting reference data to use in thematic map accuracy assessments for complex environments. This pilot study analyzed 377.57 ha of New England forests, over six University of New Hampshire woodland properties, to compare the similarity between UAS-derived orthomosaic samples and ground-based continuous forest inventory (CFI) plot classifications of deciduous, mixed, and coniferous forest cover types. Using an eBee Plus fixed-wing UAS, 9173 images were acquired and used to create six comprehensive orthomosaics. Agreement between our UAS orthomosaics and ground-based sampling forest compositions reached 71.43% for pixel-based classification and 85.71% for object-based classification reference data methods. Despite several documented sources of uncertainty or error, this research demonstrated that UAS are capable of highly efficient and effective thematic map accuracy assessment reference data collection. As UAS hardware, software, and implementation policies continue to evolve, the potential to meet the challenges of accurate and timely reference data collection will only increase.

Keywords: Unmanned Aerial Systems (UAS); structure from motion (SfM); Unmanned Aerial Vehicles (UAV); Photogrammetry; Thematic Mapping; Accuracy Assessment; Reference Data; Forest Sampling; Remote Sensing

---

1. Introduction

Growing dissidence over the causes and impacts of environmental change in the modern era has forced an ever-increasing need for data accuracy and certainty. Studied patterns of global change such as habitat augmentation, loss of biodiversity, invasive species spread, and other systems imbalances have designated humans as a ubiquitous disturbance for the natural world, leading to the current 'Anthropocene' era [1–3]. Degrading natural systems also causes noted pressures on human economies and quality of life through diminished potential for ecosystem services [4]. These services include life sustaining functions such as nutrient regulation, primary production products in agriculture and forestry, water quality management, and disease mitigation [1,2,5]. Modeling natural systems requires us to undergo the inherently difficult task of finding representative characteristics. Forested landscapes comprising high compositional and structural diversity (i.e., complexity), such as those in the Northeastern United States, further impede these efforts [6]. In many cases, land cover allows us this ability to represent fundamental constructs of the earth's surface [7]. We can then employ remote sensing as a tool to collect land cover data at scales sufficient to overcome environmental issues [8–10].

Remote sensing provides the leading source of land use and land cover data, supported by its scales of coverage, adaptability, and prolific modifications [7,11,12]. The classification of remote sensed imagery traditionally referred to as thematic mapping, labels objects and features in defined groups based on the relationship of their attributes [13,14]. This process incorporates characteristics reflected within the source imagery and motivations of the project, to recognize both natural and artificial patterns and increase our ability to make informed decisions [13,15,16].

In the digital age, the process of image classification has most often been performed on a per pixel basis. Pixel-based classification (PBC) algorithms utilize spectral reflectance values to assign class labels based on specified ranges. More refined classification techniques have also been formed to integrate data such as texture, terrain, and observed patterns based on expert knowledge [17–19].

Technologies have recently advanced to allow users a more holistic, human vision matching, approach to image analysis in the form of object-based image analysis (OBIA). Object-based classification (OBC) techniques work beyond individual pixels to distinguish image objects (i.e., polygons, areas, or features), applying additional data parameters to each individual unit [10,20,21]. OBC methods can also benefit users by reducing the noise found in land cover classifications at high spatial resolution using class-defining thresholds of spectral variability and area [22]. The specific needs of the project and the characteristics of the remote sensing data help guide the decision between which classification method would be most appropriate for creating a desired thematic layer [15,23].

Outside of the progression of classification algorithms, novel remote sensing and computer vision technologies have inspired new developments in high-resolution three-dimensional (3D) and digital planimetric modeling. Photogrammetric principles have been applied to simultaneously correct for sensor tilt, topographic displacement in the scene, relief displacement, and even lens geometric distortions [24,25]. To facilitate this process, Structure from Motion (SfM) software packages isolate and match image tie points (i.e., keypoints) within high-resolution images with sizeable latitudinal and longitudinal overlap to form 3D photogrammetric point clouds and orthomosaic models [25–27]. Techniques for accurate and effective SfM modeling have been refined, even in complex natural environments, to expand the value of these products [28–30].

The appropriate use of these emergent remote sensing data products establishes a need for understanding their accuracy and sources of error. Validating data quality is a necessary step for incorporating conclusions drawn from remote sensing within the decision-making process. Spatial data accuracy is an aggregation of two distinct characteristics: positional accuracy and thematic accuracy [10]. Positional accuracy is the difference in locational agreement between a remotely sensed data layer and known ground points, calculated through Root Mean Square Error (RMSE) [31]. Thematic accuracy expresses a more complex measure of error, evaluating the agreement for the specific labels, attributes, or characteristics between what is on the ground and the spatial data product, typically in the form of an error matrix [10].

The immense costs and difficulty of validating mapping projects have brought about several historic iterations of methods for quantitatively evaluating thematic accuracy [10]. Being once an afterthought, the assessment of thematic accuracy has matured from a visual, qualitative process into a multivariate evaluation of site-specific agreement [10]. Site-specific thematic map accuracy assessments utilize an error matrix (i.e., contingency table or confusion matrix) to evaluate individual class accuracies and the relations among sources of uncertainty [23,32]. While positional accuracy holds regulated standards for accuracy tolerance, thematic mapping projects must establish their own thresholds for amount and types of justifiable uncertainty. Within thematic accuracy two forms of error exist: errors of commission (i.e., user's accuracy) and errors of omission (i.e., producer's accuracy) [33]. Commission errors represent the user's ability to accurately classify ground characteristics [10]. Omission errors assesses if the known ground reference points have been accurately captured by the thematic layer for each class [33]. For most uses, commission errors are favored because the false addition of area to classes of interest is of less consequence than erroneously missing critical features [10]. The error matrix presents a robust quantitative analysis tool for assessing thematic map accuracy of classified maps created through both pixel-based and object-based classification methods.

Collecting reference data, whether using higher-resolution remotely sensed data, ground sampling, or previously produced sources, must be based on a sound statistical sample design. Ground sampling stands out as the most common reference data collection procedure. However, such methods generally come with an inherent greater associated cost. During the classification process, reference data can be used for two distinct purposes, depending on the applied classification algorithm. Reference data can be used to train the classification (training data), generating the decision tree ruleset which forms the thematic layer. Secondly, reference data are used as the source of validation (validation data) during the accuracy assessment. These two forms of reference data must remain independent to ensure the process is statistically valid [10].

There are also multiple methods for collecting ground reference data, such as: visual interpretation of an area, GPS locational confirmation, or full-record data sampling with precise positioning. The procedures of several professional and scientific fields have been adopted to promote the objective and efficient collection of reference data. Forest mensuration provides such a foundation for obtaining quantifiable information in forested landscapes, with systematic procedures that can mitigate biases and inaccuracies of sampling [34,35]. For many decades now, forest mensuration (i.e., biometrics) has provided the most accurate and precise observations of natural characteristics through the use of mathematical principles and field-tested tools [34–36]. To observe long-term or large area trends in forest environments, systematic Continuous Forest Inventory (CFI) plot networks have been established. Many national agencies (e.g., the U.S. Forest Service) have such a sampling design (e.g., Forest Inventory and Analysis (FIA) Program) for monitoring large land areas in a proficient manner [37]. Despite these sampling designs for efficient and effective reference data collection, the overwhelming costs of preforming a statistically valid accuracy assessment is a considerable limitation for most projects [10,23].

The maturation of remote sensing technologies in the 21st century has brought with it the practicality of widespread Unmanned Aerial Systems (UAS) applications. This low-cost and flexible platform generates on-demand, high-resolution products to meet the needs of society [38,39]. UAS represent an interconnected system of hardware and software technologies managed by a remote pilot in command [30,40]. Progressing from mechanical contraptions, UAS now assimilate microcomputer technologies that allow them to operate for forestry sampling [29,41], physical geography surveys [42], rangeland mapping [43], humanitarian aid [44], precision agriculture [45], and many other applications [12,39,46].

The added potential of the UAS platform has supported a wide diversity of data collection initiatives. UAS-SfM products provide analytical context beyond that of traditional raw imagery, with products including photogrammetric point clouds, Digital Surface Models (DSM), and planimetric (or orthomosaic) surfaces. While it is becoming increasingly common to use high-spatial resolution satellite imagery for reference data to assess maps generated from medium to coarse resolution imagery, UAS provide a new opportunity at even higher spatial resolutions. To properly apply the practice of using high-resolution remote sensing imagery as a source of validation data [36,47,48], our research focuses on if UAS provide the potential for collecting thematic map accuracy assessment reference data of a necessary quality and operational efficiency to endorse their use. To do this, we evaluated the agreement between the UAS-collected samples and the ground-based CFI plot composition. Specifically, this pilot study investigated if UAS are capable of effectively and efficiently collecting reference data for use in assessing the accuracy of thematic maps created from either a (1) pixel-based or (2) object-based classification approach.

2. Materials and Methods

This research conducted surveys of six woodland properties comprising 522.85 ha of land, 377.57 ha of which were forest cover, in Southeastern New Hampshire (Figure 1). The University of New Hampshire (UNH) owns and manages these six properties, as well as many others, to maintain research integrity for natural communities [49]. These properties contain a wide diversity of structural and compositional diversity, ranging in size from 17 ha to 94.7 ha of forested land cover. Each property also contains a network of CFI plots for measuring landscape scale forest characteristics over time.

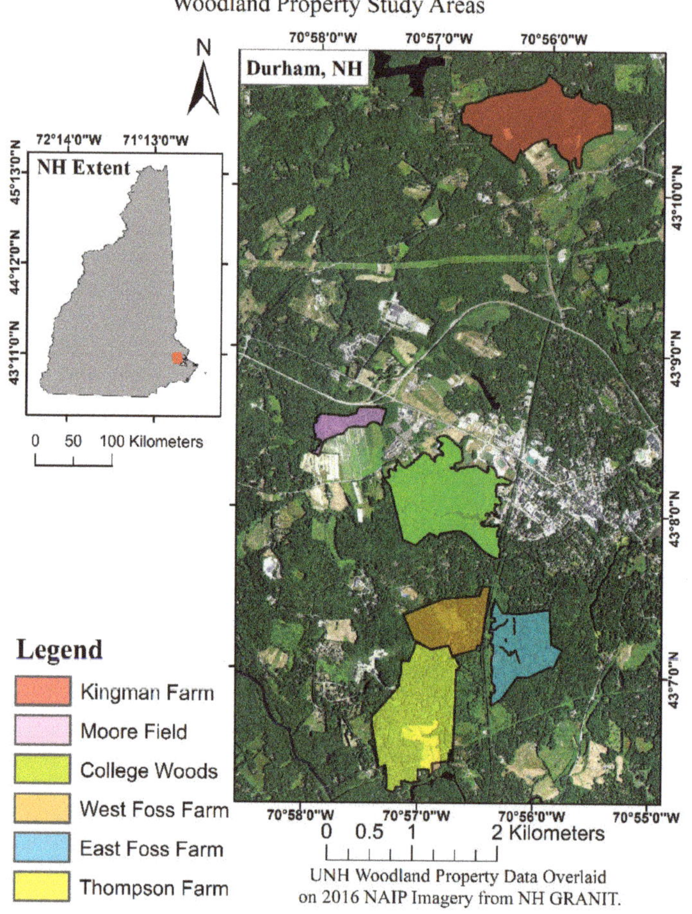

Figure 1. Woodland property boundaries for the six study areas. From North to South (with total area): Kingman Farm (135.17 ha), Moore Field (47.76 ha), College Woods (101.17 ha), West Foss Farm (52.27 ha), East Foss Farm (62.32 ha), and Thompson Farm (118.17 ha).

The systematic network of CFI ground sampling plots was established for each of the six woodland properties to estimate landscape level biophysical properties. These plot networks are sampled on a regular interval, not to exceed 10 years in reoccurrence. Kingman Farm presents the oldest data (10 years since previous sampling) and East Foss Farm, West Foss Farm, and Moore Field each being sampled most recently in 2014. CFI plots were located at one plot per hectare (Figure 2), corresponding to the minimum management unit size. Each plot location used an angle-wedge prism sampling protocol to identify the individual trees to be included in the measurement at that location. Those trees meeting the optical displacement threshold (i.e., "in" the plot) were then measured for diameter at breast height (dbh), a species presence count, and the tree species itself, through horizontal point sampling guidelines [35]. Prism sampling formed variable radius plots in relation to the basal area factor (BAF) applied. The proportional representation of species under this method is not unbiased with the basal area of the species with a larger dbh being overestimated, while those with smaller dbh are underestimated. Since photo interpretation of the plots is also performed from above this bias tends to hold here as well since the largest canopy trees are the ones most viewed. Therefore, the use of this sampling method is effective here.

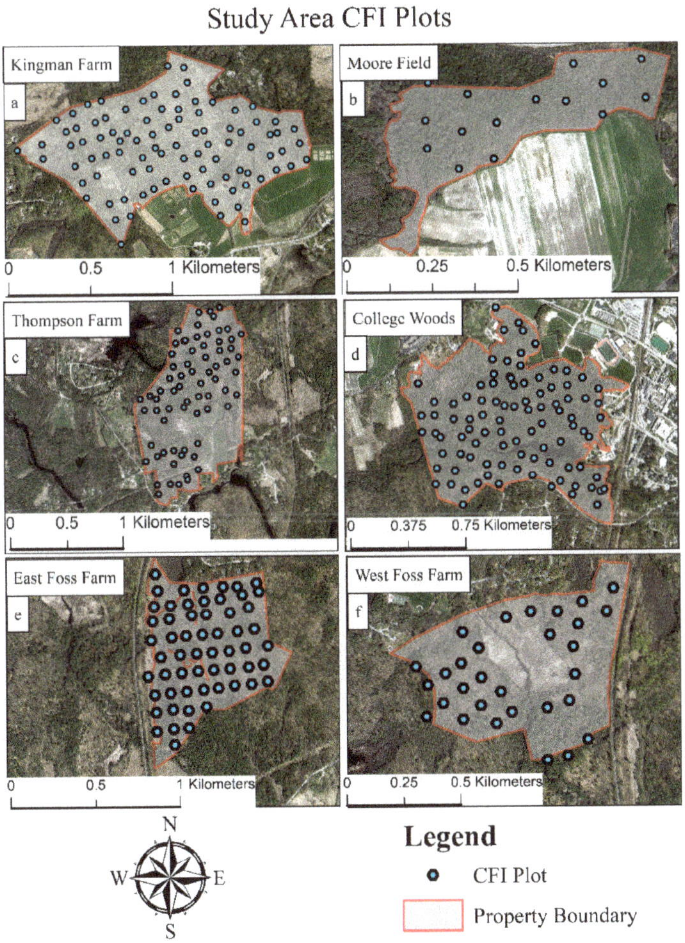

Figure 2. Woodland property continuous forest inventory (CFI) plot networks totaling 354 horizontal point sampling plots over 377.57 ha of forested land. Pictured are (Top left to bottom right): (**a**) Kingman Farm, (**b**) Moore Field, (**c**) Thompson Farm, (**d**) College Woods, (**e**) East Foss Farm, and (**f**) West Foss Farm.

The UNH Office of Woodland and Natural Areas forest technicians used the regionally recommended BAF 4.59 m^2 (or 20 ft^2) prism [50]. Additionally, a nested plot "Big BAF" sampling integration applied a BAF 17.2176 m^2 (or 75 ft^2) prism to identify a subset of trees for expanded measurements. These 'measure' trees for had their height, bearing from plot center, distance from plot center, crown dimensions, and number of silvicultural logs present recorded.

Basal area was used to characterize species distributions and proportions throughout the woodland properties [34,35]. For our study, this meant quantifying the percentage of coniferous species basal area comprising each sample. For the six observed study areas a total of 31 tree species were observed (Table S1). Instead of a species specific classification, our analysis centered on the conventional Deciduous Forest, Mixed Forest, and Coniferous Forest partitioning defined by Justice et al., [5] and MacLean et al., [6]. Here we used the Anderson et al., [7] classification scheme definition for forests, being any area with 10 percent or greater aerial tree-crown density, which has the ability to produce lumber, and influences either the climate or hydrologic regime. From this scheme we defined:

- "Coniferous" as any land surface dominated by large forest vegetation species, and managed as such, comprising an overstory canopy with a greater than or equal to 65% basal area per unit area coniferous species composition
- "Mixed Forest" being any land surface dominated by large forest vegetation species, and managed as such, comprising an overstory canopy, which is less than 65% and greater than 25% basal area per unit area coniferous species in composition.
- "Deciduous", any land surface dominated by large forest vegetation species, and managed as such, comprising an overstory canopy, which is less than or equal to 25% basal area per unit area coniferous species in composition.

The presented classification ensured that samples were mutually exclusive, totally exhaustive, hierarchical, and produced objective repeatability [7,14].

The original ground-based datasets were collected for general-purpose analysis and research and so, needed to be cleaned, recoded, and refined using R Studio, version 3.3.2 [51]. We used R Studio to isolate individual tree dbh measurements in centimeters and then compute basal area for the deciduous or coniferous species in centimeters squared. Of the original 359 CFI plots, six contained no recorded trees and were removed from the dataset, leaving 353 for analysis. Additionally, standing dead trees were removed due to the time lag between ground sampling and UAS operations. Percent coniferous composition by plot was calculated for the remaining locations based on the classification scheme.

Once classified individually as either Coniferous, Mixed, or Deciduous in composition, the CFI plot network was used to delineate forest management units (stands). Leaf-off, natural color, NH Department of Transportation imagery with a 1-foot spatial resolution (0.3 × 0.3 m) [52] provided further visual context for delineating the stand edges (Figure 3). Non-managed forests and non-forested areas were also identified and removed from the study areas.

UAS imagery was collected using the eBee Plus (SenseFly Parrot Group, Cheseaux-sur-Lausanne, Switzerland), fixed-wing platform, during June and July 2017. The SenseFly eBee Plus operated under autonomous flight missions, in eMotion3 software version 3.2.4 (senseFly SA, Cheseaux-Lausanne, Switzerland), for approximately 45 minutes per battery. This system deployed the sensor optimized for drone applications (S.O.D.A), a 20 megapixel, 1in (2.54 cm) global shutter, natural color, proprietary sensor designed for photogrammetric analysis. In total, the system weighed 1.1 kg (Figure 4).

Study Area Forest Stands

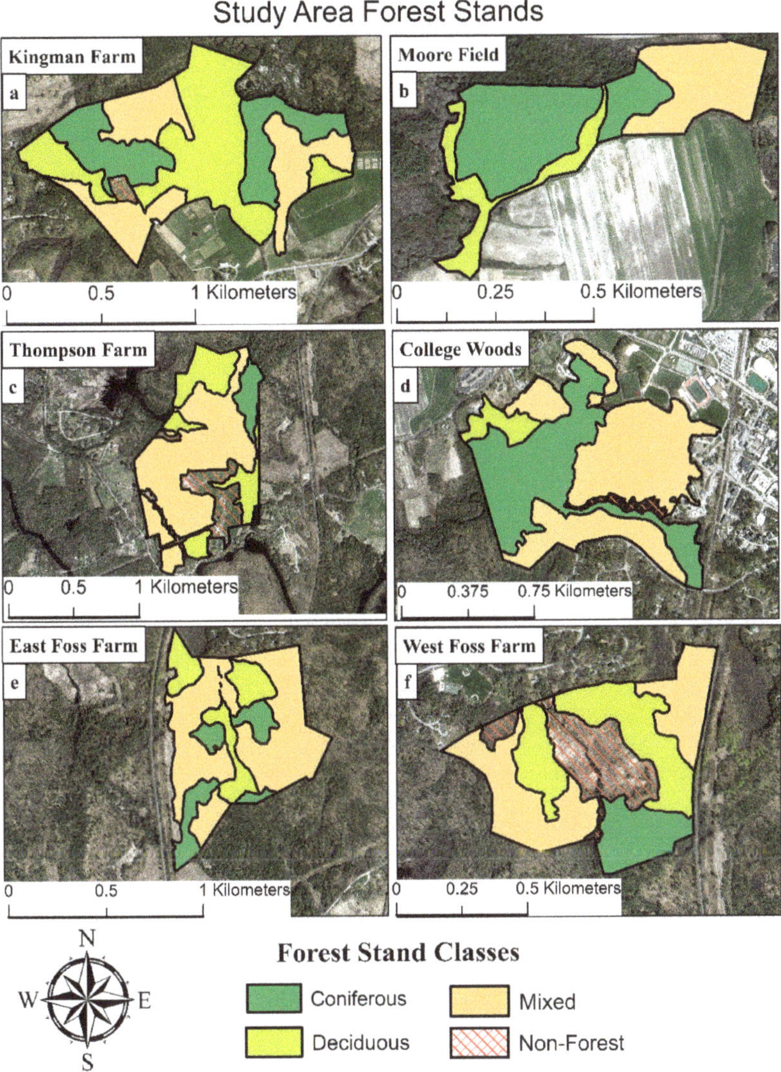

Figure 3. Ground-based forest stands digitized from CFI-plot classifications. Pictured are (Top left to bottom right): (**a**) Kingman Farm, (**b**) Moore Field, (**c**) Thompson Farm, (**d**) College Woods, (**e**) East Foss Farm, and (**f**) West Foss Farm.

UAS mission planning was designed to capture plot- and stand- level forest composition. Our team predefined mission blocks which optimized image collection while minimizing time outside of the study area. For larger properties (e.g., College Woods) up to six mission blocks were required based on legal restrictions to comprehensively image the study area. We used the maximum allowable flying height of 120 m above the forest canopy with an 85% forward overlap, and 75% side overlap for all photo missions [30,53]. This flying height was set relative to a statewide LiDAR dataset canopy height model provided by New Hampshire GRANIT [54]. Further characteristics such as optimal sun angle (e.g., around solar noon), perpendicular wind directions, and consistent cloud coverage were considered during photo missions to maintain image quality and precision [28,30].

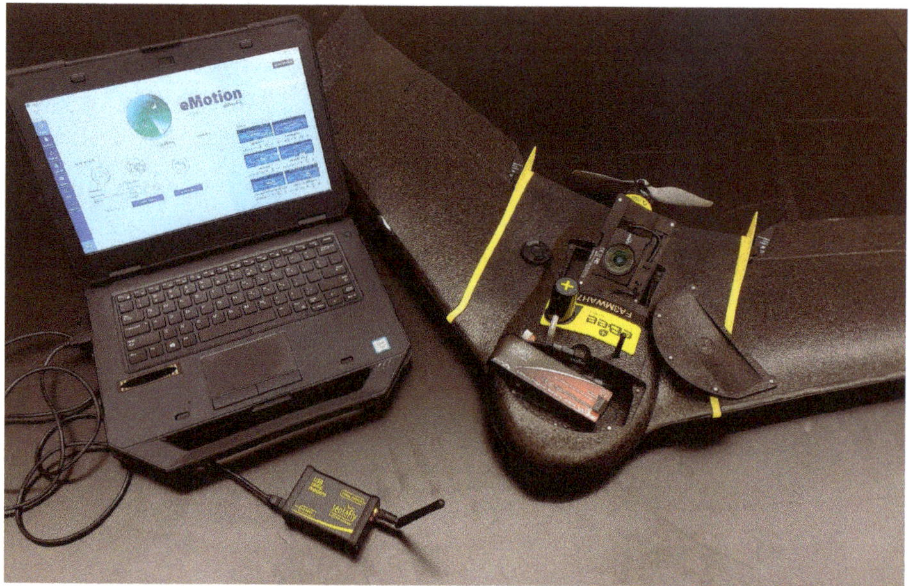

Figure 4. eBee Plus Unmanned Aerial System (UAS) with the sensor optimized for drone applications (S.O.D.A) and eMotion3 flight planning software.

Post-flight processing began with joining the spatial data contained within the onboard flight log (.bb3 or .bbx) to each individual captured image. Next, we used Agisoft PhotoScan 1.3.2 [55] for a high accuracy photo alignment, image tie point calibration, medium-dense point cloud formation, and planimetric model processing workflow [30]. For all processing, we used a Dell Precision 7910, running an Intel Xeon E5-2697 v4 (18 core) CPU, with 64 GB of RAM, and a NVIDIA Quadro M4000 graphics card. Six total orthomosaics were created.

For each classification method, UAS reference data were extracted from the respective woodland property orthomosaic. West Foss Farm was used solely for establishing training data samples to guide the photo interpretation processes. In total, there were six sampling methods for comparing the ground-based and UAS derived reference data (Table 1) (Figures S1–S6).

Table 1. Six total methods used for UAS reference data collection, between Pixel-based (PBC) and Object-based (OBC) classification approaches.

Classification Approach	
Pixel-based Classification	Object-based Classification
1. Stratified Random Distribution	3. Stratified Random, Individual Subsamples
2. CFI-plot Positionally Dependent	4. Stratified Random, Image Object Majority Agreement
	5. CFI-plot Dependent, Individual Subsamples
	6. CFI-plot Dependent, Image Object Majority Agreement

For the first pixel-based classification reference data collection method (method one), 90×90 m extents were partitioned into 30×30 m grids, and positioned at the center of each forest stand. The center 30×30 m area then acted as the effective area for visually classifying the given sample. Using an effective area in this way both precluded misregistration errors between the reference data and the thematic layer, and ensured that the classified area was fully within the designated stand boundary [10]. The second PBC reference data collection method (method two) used this same

90 × 90 m partitioned area but positionally aligned it with CFI-plot locations, avoiding overlaps with boundaries and other samples.

The first of four object-based classification reference data collection methods (method three) used a stratified random distribution for establishing a maximum number of 30 × 30 m interpretation areas (subsamples) within each forest stand. In total, 268 of these samples were created throughout 35 forest stands while remaining both spatially independent and maintaining at least two samples per forest stand. Similar to both PBC sampling methods, this and other OBC samples used 30 × 30 m effective areas for visually interpreting their classification. The second OBC reference data collection method (method four) used these previous 30 × 30 m classified areas as subsamples to represent the compositional heterogeneity at the image object (forest stand) level [5,10]. Forest stands which did not convey a clear majority, based on the subsamples, were classified based on a decision ruleset shown in Table 2.

Table 2. Decision support ruleset for forest stands (image objects) classification of split decision areas.

Class 1	Class 2	Resulting Classification
Coniferous	Mixed	Coniferous
Deciduous	Mixed	Deciduous
Coniferous	Deciduous	Mixed

For the remaining two OBC reference data collection methods, we assessed individual 30 × 30 m samples (method five) and the overall forest stand classifications (method six) by direct comparison with the CFI-plots location compositions. An internal buffer of 21.215 m (the hypotenuse of the 30 × 30 m effective area) was applied to each forest stand to eliminate CFI-plots that were subject to stand boundary overlap. This process resulted in 202 subsamples for 28 stands within the interior regions of the five classified woodland properties.

For each of the six orthomosaic sampling procedures we relied on photo interpretation for deriving their compositional cover type classification. Using a confluence of evidence within the imagery, including morphological and spatial distribution patterns, the relative abundance of coniferous and deciduous species was identified [24,56]. Supporting this process was the training data collected from West Foss Farm (Figure S7). A photo interpretation key was generated for plots with distinct compositional proportions, set at the distinctions between coniferous, deciduous, and mixed forest classes. During the visual classification process, a blind interpretation method was used so that ground data bias or location was not influential.

Error matrices were used to quantify the agreement between the UAS orthomosaic and ground-based thematic map reference data samples. Sample units for both PBC and OBC across all six approaches followed this method. These site-specific assessments reported producer's, user's, and overall accuracies for the five analyzed woodland properties [33].

3. Results

UAS imagery across the six woodland properties was used to generate six orthomosaics with a total land cover area of 398.71 ha. These UAS-SfM models represented 9173 images (Figure 5). The resulting ground sampling distances (gsd) were: Kingman Farm at 2.86 cm, Moore Field at 3.32 cm, East Foss Farm at 3.54 cm, West Foss Farm at 3.18 cm, Thompson Farm at 3.36 cm, and College Woods at 3.19 cm, for an average pixel size of 3.24 cm. The use of Agisoft PhotoScan for producing these orthomosaics does not report XY positional errors. Additional registration of the woodland areas modeled to another geospatial data layer could determine relative error.

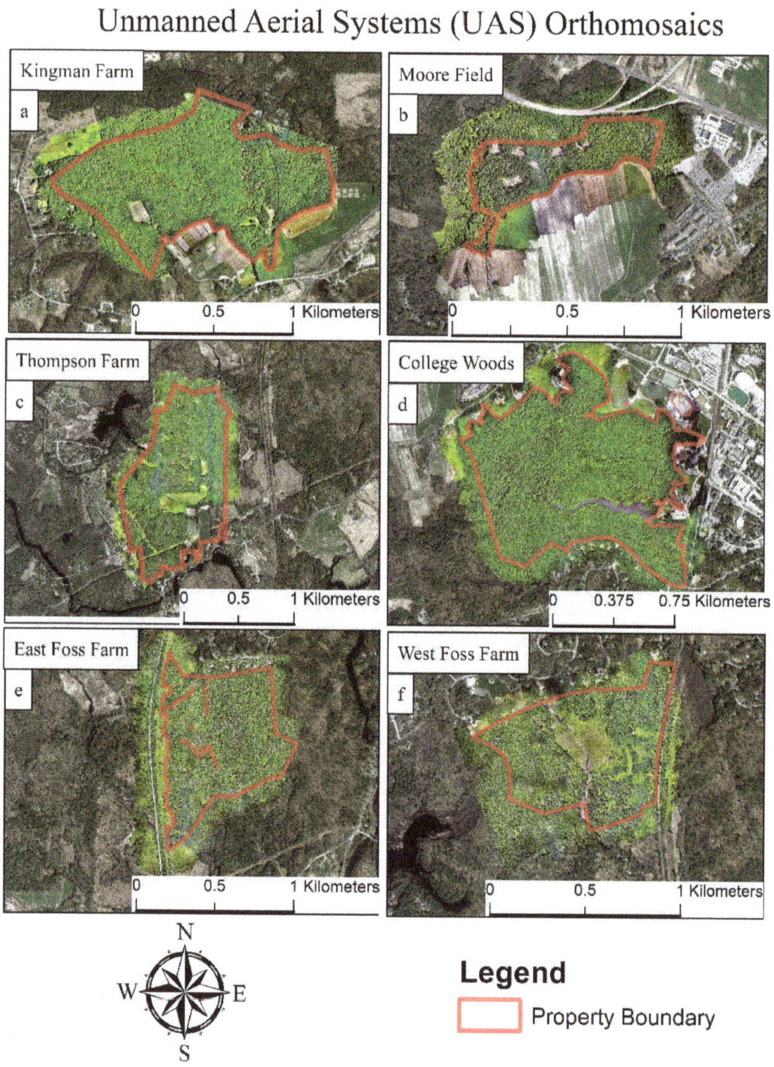

Unmanned Aerial Systems (UAS) Orthomosaics

Figure 5. UAS orthomosaics for the six woodland properties (Top left to Bottom Right): (**a**) Kingman Farm, (**b**) Moore Field, (**c**) Thompson Farm, (**d**) College Woods, (**e**) East Foss Farm, (**f**) West Foss Farm.

In our first analysis of pixel-based classification thematic map accuracy assessment reference data agreement, 29 sample units were located at the center of the forest stands. This method represented the photo interpretation potential of classifying forest stands from UAS image products. Overall agreement between ground-based and UAS-based reference data samples was 68.97% (Table 3). Producer's accuracy was highest for deciduous stands, while user's accuracy was highest for coniferous forest stands.

Table 3. Stratified random sampling PBC thematic map error matrix. Ground (reference) data are represented by the CFI plots and Unmanned Aerial Systems (UAS) data are derived from the corresponding orthomosaic.

		Ground Data				User's Accuracy
		Coniferous	Mixed	Deciduous	Total	
	Coniferous	4	1	0	5	80.0%
UAS	Mixed	3	8	2	13	61.54%
	Deciduous	0	3	8	11	72.73%
	Total	7	12	10	29	
Producer's Accuracy		57.14%	66.67%	80.0%		Overall Accuracy = 20/29 or 68.97%

For our second PBC reference data analysis, in which orthomosaic samples were registered with CFI-plot locations, 19 samples were assessed. Reference data classification agreement was 73.68% (Table 4), with both user's and producer's accuracies highest for coniferous forest stands.

Table 4. CFI plot-registered PBC thematic map error matrix. Ground (reference) data are represented by the CFI plots and Unmanned Aerial Systems (UAS) data are derived from the corresponding orthomosaic.

		Ground Data				User's Accuracy
		Coniferous	Mixed	Deciduous	Total	
	Coniferous	5	0	0	5	100%
UAS	Mixed	1	5	2	8	62.5%
	Deciduous	0	2	4	6	66.66%
	Total	6	7	6	19	
Producer's Accuracy		83.33%	71.43%	66.66%		Overall Accuracy = 14/19 or 73.68%

Four total OBC reference data error matrices were generated; two for the individual subsamples and two for the forest stands or image objects. Using the stratified random distribution for subsamples, our analysis showed an overall agreement of 63.81% between the ground-based forest stands and UAS orthomosaics across 268 samples. Producer's accuracy was highest for deciduous forests while user's accuracy was highest for mixed forest (Table 5).

Table 5. Stratified randomly distributed OBC reference data subsample error matrix. Ground (reference) data are represented by the CFI plots and Unmanned Aerial Systems (UAS) data are derived from the corresponding orthomosaic.

		Ground Data				User's Accuracy
		Coniferous	Mixed	Deciduous	Total	
	Coniferous	40	18	1	59	68.0%
UAS	Mixed	23	90	15	128	70.31%
	Deciduous	3	37	41	81	50.62%
	Total	66	145	57	268	
Producer's Accuracy		60.61%	62.07%	71.93%		Overall Accuracy = 171/268 or 63.81%

At the forest stand level, the majority agreement of the stratified randomly distributed subsamples presented a 71.43% agreement when compared to the ground-based forest stands (Table 6). For the 35 forest stands analyzed, user's accuracy was 100% for coniferous forest stands. Producer's accuracy was highest for deciduous stands at 81.82%.

Table 6. OBC sample unit thematic map error matrix for stratified randomly distributed subsamples. Ground (reference) data are represented by the CFI plots and Unmanned Aerial Systems (UAS) data are derived from the corresponding orthomosaic.

		Ground Data				User's Accuracy
		Coniferous	Mixed	Deciduous	Total	
	Coniferous	7	0	0	7	100%
UAS	Mixed	4	9	2	15	60.0%
	Deciduous	0	4	9	13	69.23%
	Total	11	13	11	35	
Producer's Accuracy		63.64%	69.23%	81.82%		Overall Accuracy = 25/35 or 71.43%

Next, UAS orthomosaic subsamples that were positionally aligned with individual CFI plots were assessed. A total of 202 samples were registered, with a 61.88% classification agreement (Table 7). User's accuracy was again highest for coniferous stands at 91.80%. Producer's accuracy for these subsamples was highest in mixed forest, with an 80.85% agreement.

Table 7. CFI plot-registered UAS orthomosaic subsample thematic map error matrix. Ground (reference) data are represented by the CFI plots and Unmanned Aerial Systems (UAS) data are derived from the corresponding orthomosaic.

		Ground Data				User's Accuracy
		Coniferous	Mixed	Deciduous	Total	
	Coniferous	56	1	4	61	91.80%
UAS	Mixed	41	38	17	96	39.58%
	Deciduous	6	8	31	45	68.89%
	Total	103	47	52	202	
Producer's Accuracy		54.37%	80.85%	59.62%		Overall Accuracy = 125/202 or 61.88%

Forest stand level classification agreement, based on the positionally registered orthomosaic samples was 85.71%. In total, 28 forest stands were assessed (Table 8). User's and producer's accuracies for all three classes varied marginally, ranging from 84.62% to 87.51%. Commission and omission error were both lowest for deciduous forest stands.

Table 8. UAS forest stand thematic map error matrix for CFI plot-registered samples. Ground (reference) data are represented by the CFI plots and Unmanned Aerial Systems (UAS) data are derived from the corresponding orthomosaic.

		Ground Data				User's Accuracy
		Coniferous	Mixed	Deciduous	Total	
	Coniferous	6	1	0	7	85.71%
UAS	Mixed	1	11	1	13	84.62%
	Deciduous	0	1	7	8	87.50%
	Total	7	13	8	28	
Producer's Accuracy		85.71%	84.62%	87.50%		Overall Accuracy = 24/28 or 85.71%

4. Discussion

This research set out to gauge whether UAS could adequately collect reference data for use in thematic map accuracy assessments, of both pixel-based and object-based classifications, for complex forest environments. To create UAS based comparative reference data samples, six independent orthomosaic models, totaling 398.71 ha of land area were formed from 9173 images (Figure 5).

The resulting average gsd was 3.24 cm. For the six comparative analyses of UAS and ground-based reference data (Table 1), 581 samples were used.

Beginning with PBC, the resulting agreement for stratified randomly distributed samples was 68.97% (Table 3). For this sampling technique, we experienced high levels of commission errors, especially between the coniferous and mixed forest types. One reason for this occurrence could have been the perceived dominance, visual bias, of the conifer canopies within the orthomosaic samples. Mixed forests experienced the greatest mischaracterization here. The CFI plot-registered PBC method generated a slightly higher overall accuracy at 73.68% (Table 4). The mixed forest samples still posed issues for classification. Coniferous samples however, showed much improved agreement with ground-based classifications.

Next, we looked at the object-based classification reference data samples. Stratified randomly distributed subsamples had an agreement of 63.81% (Table 5). While at the forest stand level agreement to the ground-based composition was 71.43% (Table 6). As before, mixed forest samples showed the highest degree of error. CFI plot-registered OBC subsamples have a 61.88% agreement (Table 7). For forest stand classifications based on these plot-registered subsamples, agreement was 85.71% (Table 8). Mixed forests once again led to large amounts of both commission and omission error. Other than OBC subsample assessments, our results showed a continuously lower accuracy for the stratified randomly distributed techniques. The patchwork composition of the New England forest landscape could be a major reason for this difficulty.

As part of our analysis we wanted to understand the sources of intrinsic uncertainty for UAS reference data collection [10,18]. The compositional and structural complexity, although not to the degree of tropical forests, made working with even the three classes difficult. Visual interpretation was especially labored by this heterogeneity. To aid the interpretation process, branching patterns and species distribution trends were used [24,56]. All visual based classification was performed by the same interpreter, who has significant experience in remote sensing photo interpretation as well as local knowledge of the area. Another source of error could have been from setting fixed areas for UAS-based reference data samples while the CFI plots established variable radius areas [35]. Our 30x30m effective areas looked to capture the majority of ground measured trees, providing snapshots of similarly sized sampling areas. Lastly, there were possible sources of error stemming from the CFI plot ground sampling procedures. Some woodlots, such as Kingman Farm, were sampled up to 10 years ago. Slight changes in composition could have occurred. Also, GPS positional error for the CFI plots was a considerable concern given the dense forest canopies. Error in GPS locations were minimized by removing points close to stand boundaries and by using pixel clusters when possible.

One of the first difficulties encountered in this project was in the logistics of flight planning. While most practitioners may strive for flight line orientation in a cardinal direction, we were limited at some locations due to FAA rules and abutting private properties [30,57]. As stated in the methods, UAS training missions and previously researched advice were used to guide comprehensive coverage of the woodland properties [28]. A second difficulty in UAS reference data collection was that even with a sampling area of 377.57 ha, the minimum statistically valid sample size for a thematic mapping accuracy assessment was not reached [10]. Forest stand structure and arrangement limited the number of samples for most assessments to below the recommended samples size of approximately 30 per class. A considerably larger, preferably continuous, forested land area would be needed to generate a sufficient sampling design. Limited sample sizes also brought into perspective the restriction from a more complex classification scheme. Although some remote sensing studies have performed to a species-specific classification, Justice et al., [5] and MacLean et al., [6] have both shown that a broader, three class, scheme has potential for understanding local forest composition.

Despite the still progressing nature of UAS data collection applications, this study has made the potential for cost reductions apparent. The volume of data collected and processed in only a few weeks opened the door for potential future research in digital image processing and computer vision. Automated classification processing, multiresolution segmentation [20], or machine learning were a consideration but could not be implemented in this study. A continuing goal is to integrate the added context of the digital surface model (DSM), texture, and multispectral image properties into automated forest classifications. We hope that in future studies more precise ground data can be collected, to alleviate the positional registration error and help match exact trees. Additionally, broader analyses should be conducted to establish a comparison for UAS-based reference data to other forms of ground-based sampling protocols (e.g., FIA clustered sampling or fixed-area plots). Lastly, multi-temporal imagery could benefit all forms of UAS classification and should be studied further.

In well under a months' time, this pilot study collected nearly 400 ha of forest land cover data to a reasonable accuracy. With added expert knowledge-driven interpretation or decreased landscape heterogeneity, this platform could prove to be a significant benefit to forested area research and management. Dense photogrammetric point clouds and ultra-high-resolution orthomosaic models were obtained, with the possibility of incorporating multispectral imagery in the future. These ultra-high resolution products have the potential now to provide an accessible alternative to reference data collected using high-spatial resolution satellite-based imagery. For the objective of collecting reference data which can train and validate environmental models, it must be remembered that reference data itself is not without intrinsic error [58]. As hardware and software technologies continue to improve, the efficiency and effectiveness of these methods will continue to grow [39]. UAS positional accuracy assessment products are gaining momentum [12,59,60]. Providing examples to the benefits of UAS should also support further legislative reform, better matching the needs of practitioners. FAA RPIC guidelines remain a sizeable limitation for UAS mapping of continuous, remote, or structurally complex areas [39,57,61]. We should also remember that these technologies should be used to augment and enhance data collection initiatives, and not replace the human element in sampling.

5. Conclusions

The collection of reference data for the training and validation of earth systems models bares considerable costs yet remains an essential component for prudent decision-making. The objectives of this pilot study were to determine if the application of UAS could enhance or support the collection of thematic map accuracy assessment reference data for both pixel-based and object-based classification of complex forests. Comparative analyses quantified the level of agreement between ground-based CFI plot compositions and that of UAS-SfM orthomosaic samples. Despite diminished agreement from mixed forest areas, PBC showed 68.97% agreement for stratified randomly distributed samples and 73.86% for CFI plot-registered samples. For OBC classifications, forest stands reached 71.43% agreement for stratified randomly distributed samples and 85.71% for CFI plot-registered samples. Our results demonstrated the ability to comprehensively map nearly 400 ha of forest area, using a UAS, in only a few weeks' time. They also showed the significant benefit that could be gained from deploying UAS to capture forest landscape composition. Low sample sizes, positional error in the CFI plot measurements, and photo interpretation insensitivity could have led to heightened commission and omission errors. Along with these sources of uncertainty, our results should be considered with the understanding that all reference data has intrinsic error and that UAS are not presented to be total replacements of in situ data collection initiatives. The continual advancement of the platform however, should be the basis for investigating their use in a greater number of environments, for the comparison to more varied ground-based reference data frameworks, and with the inclusion of more technologically advanced classification procedures.

Supplementary Materials: The following are available online at http://www.mdpi.com/1999-4907/10/1/24/s1, Figure S1: Pixel-based classification, Stratified Random Distribution sampling design diagram; Figure S2: Pixel-based classification, CFI-plot Positionally Dependent sampling design diagram; Figure S3: Object-based Classification, Stratified Random, Individual Subsamples sampling design diagram; Figure S4: Object-based Classification, Stratified Random Image Object Majority Agreement sampling design diagram; Figure S5: Object-based Classification, CFI-plot Dependent Individual Subsamples sampling design diagram; Figure S6: Object-based Classification, CFI-plot Dependent, Image Objects Majority Agreement sampling design diagram.

Author Contributions: B.T.F. and R.G.C. conceived and designed the experiments; B.T.F. performed the experiments and analyzed the data with guidance from R.G.C.; R.G.C. provided the materials and analysis tools; B.T.F. wrote the paper; R.G.C. revised the paper and manuscript.

Funding: Partial funding was provided by the New Hampshire Agricultural Experiment Station. The Scientific Contribution Number is 2798. This work was supported by the USDA National Institute of Food and Agriculture McIntire Stennis Project #NH00077-M (Accession #1002519).

Acknowledgments: These analyses utilized processing within the Agisoft PhotoScan Software Package with statistical outputs generated from its results. All UAS operations were conducted on University of New Hampshire woodland properties with permission from local authorities and under the direct supervision of pilots holding Part 107 Remote Pilot in command licenses.

Conflicts of Interest: The authors declare no conflict of interest.

References

1. Chapin, F.S., III; Zavaleta, E.S.; Eviner, V.T.; Naylor, R.L.; Vitousek, P.M.; Reynolds, H.L.; Hooper, D.U.; Lavorel, S.; Sala, O.E.; Hobbie, S.E.; et al. Consequences of changing biodiversity. *Nature* **2000**, *405*, 234. [CrossRef] [PubMed]

2. Pejchar, L.; Mooney, H.A. Invasive species, ecosystem services and human well-being. *Trends Ecol. Evol.* **2009**, *24*, 497–504. [CrossRef] [PubMed]

3. McGill, B.J.; Dornelas, M.; Gotelli, N.J.; Magurran, A.E. Fifteen forms of biodiversity trend in the Anthropocene. *Trends Ecol. Evol.* **2015**, *30*, 104–113. [CrossRef] [PubMed]

4. Kareiva, P.; Marvier, M. *Conservation Science: Balancing the Needs of People and Nature*, 1st ed.; Roberts and Company Publishing: Greenwood Village, CO, USA, 2011; p. 543, ISBN 1936221063, 9781936221066.

5. MacLean, M.G.; Campbell, M.J.; Maynard, D.S.; Ducey, M.J.; Congalton, R.G. Requirements for labeling forest polygons in an object-based image analysis classification. *Int. J. Remote Sens.* **2013**, *34*, 2531–2547. [CrossRef]

6. Justice, D.; Deely, A.K.; Rubin, F. *Land Cover and Land Use Classification for the State of New Hampshire, 1996–2001*; ORNL DAAC: Oak Ridge, TN, USA, 2016. [CrossRef]

7. Anderson, J.R.; Hardy, J.T.; Roach, J.T.; Witmer, R.E. A Land Use and Land Cover Classification System for Use with Remote Sensor Data. Geological Survey Professional Paper 964; 1976. Available online: https://landcover.usgs.gov/pdf/anderson.pdf (accessed on July 2017).

8. Field, C.B.; Randerson, J.T.; Malmström, C.M. Global net primary production: Combining ecology and remote sensing. *Remote Sens. Environ.* **1995**, *51*, 74–88. [CrossRef]

9. Ford, E.D. *Scientific Method for Ecological Research*, 1st ed.; Cambridge University Press: Cambridge, UK, 2000; p. 564, ISBN 0521669731.

10. Congalton, R.G.; Green, K. *Assessing the Accuracy of Remotely Sensed Data: Principles and Practices*, 2nd ed.; CRC Press, Taylor & Francis Group: Boca Raton, FL, USA, 2009; p. 183, ISBN 978-1-4200-5512-2.

11. Turner, M.G. Landscape Ecology: What Is the State of the Science? *Annu. Rev. Ecol. Evol. Syst.* **2005**, *36*, 319–344. [CrossRef]

12. Whitehead, K.; Hugenholtz, C.H. Remote sensing of the environment with small unmanned aircraft systems (UASs), part 1: A review of progress and challenges. *J. Unmanned Veh. Syst.* **2014**, *02*, 69–85. [CrossRef]

13. Sokal, R.R. Classification: Purposes, Principles, Progress, Prospects. *Science* **1974**, *185*, 1115–1123. [CrossRef]

14. Jensen, J.R. *Introductory Digital Image Processing: A Remote Sensing Perspective*, 4th ed.; Pearson Education Inc.: Glenview, IL, USA, 2016; p. 544, ISBN 978-0134058160.

15. Pugh, S.A.; Congalton, R.G. Applying Spatial Autocorrelation Analysis to Evaluate Error in New England Forest-Cover-Type maps derived from Landsat Thematic Mapper Data. *Photogramm. Eng. Remote Sens.* **2001**, *67*, 613–620.

16. Kerr, J.T.; Ostrovsky, M. From space to species: Ecological applications for remote sensing. *Trends Ecol. Evol.* **2003**, *18*, 299–305. [CrossRef]

17. Harris, P.M.; Ventura, S.J. The integration of geographic data with remotely sensed imagery to improve classification in an urban area. *Photogramm. Eng. Remote Sens.* **1995**, *61*, 993–998.

18. Lu, D.; Weng, Q. A survey of image classification methods and techniques for improving classification performance. *Int. J. Remote Sens.* **2007**, *28*, 823–870. [CrossRef]

19. Caridade, C.M.R.; Marçal, A.R.S.; Mendonça, T. The use of texture for image classification of black & white air photographs. *Int. J. Remote Sens.* **2008**, *29*, 593–607. [CrossRef]

20. Blaschke, T. Object based image analysis for remote sensing. *ISPRS J. Photogramm. Remote Sens.* **2010**, *65*, 2–16. [CrossRef]

21. Radoux, J.; Bogaert, P.; Fasbender, D.; Defourny, P. Thematic accuracy assessment of geographic object-based image classification. *Int. J. Geogr. Inf. Sci.* **2011**, *25*, 895–911. [CrossRef]

22. Robertson, L.D.; King, D.J. Comparison of pixel- and object-based classification in land cover change mapping. *Int. J. Remote Sens.* **2011**, *32*, 1505–1529. [CrossRef]

23. Foody, G.M. Status of land cover classification accuracy assessment. *Remote Sens. Environ.* **2002**, *80*, 185–201. [CrossRef]

24. Avery, T.E.; Berlin, G.L. *Interpretation of Aerial Photographs*, 4th ed.; Burgess Publishing Company: Minneapolis, MN, USA, 1985; p. 554, ISBN 978-0808700968.

25. Westoby, M.J.; Brasington, J.; Glasser, N.F.; Hambrey, M.J.; Reynolds, J.M. 'Structure-from-Motion' photogrammetry: A low-cost, effective tool for geoscience applications. *Geomorphology* **2012**, *179*, 300–314. [CrossRef]

26. Nex, F.; Remondino, F. UAV for 3D mapping applications: A review. *Appl. Geomat.* **2014**, *6*, 1–15. [CrossRef]

27. Micheletti, N.; Chandler, J.H.; Lane, S.N. Structure from motion (SFM) photogrammetry. In *Geomorphological Techniques*; Cook, S.J., Clarke, L.E., Nield, J.M., Eds.; British Society for Geomorphology: London, UK, 2012; Chapter 2, Section 2.2, ISSN 2047-0371.

28. Dandois, J.P.; Olano, M.; Ellis, E.C. Optimal Altitude, Overlap, and Weather Conditions for Computer Vision UAV Estimates of Forest Structure. *Remote Sens.* **2015**, *7*, 13895–13920. [CrossRef]

29. Mikita, T.; Janata, P.; Surový, P. Forest Stand Inventory Based on Combined Aerial and Terrestrial Close-Range Photogrammetry. *Forests* **2016**, *7*, 165. [CrossRef]

30. Fraser, B.T.; Congalton, R.G. Issues in Unmanned Aerial Systems (UAS) Data Collection of Complex Forest Environments. *Remote Sens.* **2018**, *10*, 908. [CrossRef]

31. Bolstad, P. *GIS Fundamentals: A First Text on Geographic Information Systems*, 4th ed.; Eider Press: White Bear Lake, MN, USA, 2012; 688p, ISBN 978-0971764736.

32. Congalton, R.G. A review of assessing the accuracy of classifications of remotely sensed data. *Remote Sens. Environ.* **1991**, *37*, 35–46. [CrossRef]

33. Story, M.; Congalton, R.G. Accuracy Assessment: A User's Perspective. *Photogramm. Eng. Remote Sens.* **1986**, *52*, 397–399.

34. Husch, B.; Miller, C.I.; Beers, T.W. *Forest Mensuration*, 2nd ed.; Ronald Press Company: New York, NY, USA, 1972.

35. Kershaw, J.A.; Ducey, M.J.; Beers, T.W.; Husch, B. *Forest Mensuration*, 5th ed.; John Wiley and Sons: Hoboken, NJ, USA, 2016; 632p, ISBN 9781118902035.

36. Spurr, S.H. *Forest Inventory*; Ronald Press Company: New York, NY, USA, 1952; 476p.

37. Smith, W.B. Forest inventory and analysis: A national inventory and monitoring program. *Environ. Pollut.* **2002**, *116*, S233–S242. [CrossRef]

38. Marshall, D.M.; Barnhart, R.K.; Shappee, E.; Most, M. *Introduction to Unmanned Aircraft Systems*, 2nd ed.; CRC Press: Boca Raton, FL, USA, 2016; p. 233, ISBN 978-1482263930.

39. Cummings, A.R.; McKee, A.; Kulkarni, K.; Markandey, N. The Rise of UAVs. *Photogramm. Eng. Remote Sens.* **2017**, *83*, 317–325. [CrossRef]

40. Colomina, I.; Molina, P. Unmanned aerial systems for photogrammetry and remote sensing: A review. *ISPRS J. Photogramm. Remote Sens.* **2014**, *92*, 79–97. [CrossRef]

41. Tang, L.; Shao, G. Drone remote sensing for forestry research and practices. *J. For. Res.* **2015**, *26*, 791–797. [CrossRef]

42. Smith, M.W.; Carrivick, J.L.; Quincey, D.J. Structure from motion photogrammetry in physical geography. *Prog. Phys. Geogr.* **2016**, *40*, 247–275. [CrossRef]

43. Laliberte, A.S.; Goforth, M.A.; Steele, C.M.; Rango, A. Multispectral Remote Sensing from Unmanned Aircraft: Image Processing Workflows and Applications for Rangeland Environments. *Remote Sens.* **2011**, *3*, 2529–2551. [CrossRef]

44. Kakaes, K.; Greenwood, F.; Lippincott, M.; Dosemagen, S.; Meier, P.; Wich, S. *Drones and Aerial Observation: New Technologies for Property Rights, Human Rights, and Global Development a Primer*; New America: Washington, DC, USA, 22 July 2015; p. 104.

45. Primicerio, J.; Di Gennaro, S.F.; Fiorillo, E.; Genesio, L.; Lugato, E.; Matese, A.; Vaccari, F.P. A flexible unmanned aerial vehicle for precision agriculture. *Precis. Agric.* **2012**, *13*, 517–523. [CrossRef]

46. Watts, A.C.; Ambrosia, V.G.; Hinkley, E.A. Unmanned Aircraft Systems in Remote Sensing and Scientific Research: Classification and Considerations of Use. *Remote Sens.* **2012**, *4*, 1671–1692. [CrossRef]

47. McRoberts, R.E.; Tomppo, E.O. Remote sensing support for national forest inventories. *Remote Sens. Environ.* **2007**, *110*, 412–419. [CrossRef]

48. Yadav, K.; Congalton, R.G. Issues with Large Area Thematic Accuracy Assessment for Mapping Cropland Extent: A Tale of Three Continents. *Remote Sens.* **2018**, *10*, 53. [CrossRef]

49. University of New Hampshire, Office of Woodlands and Natural Areas, General Information. Available online: https://colsa.unh.edu/woodlands/general-information (accessed on 24 July 2017).

50. Ducey, M.J. Pre-Cruise Planning. In *Workshop Proceedings: Forest Measurements for Natural Resource Professionals*; Natural Resource Network: Connecting Research, Teaching and Outreach; University of New Hampshire Cooperative Extension: Durham, NH, USA, October 2001.

51. RStudio Team. *RStudio: Integrated Development for R*; RStudio Inc.: Boston, MA, USA, 2016. Available online: http://www.rstudio.com/ (accessed on July 2017).

52. New Hampshire GRANIT: New Hampshire Statewide GIS Clearinghouse, 2015, Aerial Photography. Available online: http://granit.unh.edu/resourcelibrary/specialtopics/2015aerialphotography/index.html (accessed on June 2017).

53. *Pix4DMapper User Manual Version 3.2*; Pix4D SA: Lausanne, Switzerland, 2017.

54. New Hampshire GRANIT LiDAR Distribution Site. Available online: http://lidar.unh.edu/map/ (accessed on 5 June 2017).

55. Agisoft PhotoScan Professional Edition. Version 1.3.2 Software. Available online: http://www.agisoft.com/downloads/installer/ (accessed on July 2017).

56. Avery, T.E. *Interpretation of Aerial Photographs*, 3rd ed.; Burgess Publishing Company: Minneapolis, MN, USA, 1977; p. 393, ISBN1 0808701304, ISBN2 9780808701309.

57. Federal Aviation Administration, Fact Sheet-Small Unmanned Aircraft Regulations (Part 107). Available online: https://www.faa.gov/news/fact_sheets/news_story.cfm?newsId=20516 (accessed on July 2017).

58. Fitzpatrick-Linz, K. Comparison on sampling procedures and data analysis for a land-use and land-cover map. *Photogramm. Eng. Remote Sens.* **1981**, *47*, 343–351.

59. Naumann, M.; Geist, M.; Bill, R.; Niemeyer, F.; Grenzdorffer, G. Comparison on sampling procedures and data analysis for a land-use and land-cover map. In Proceedings of the International Archives of the Photogrammetry, Remote Sensing and Spatial Information Sciences 2013, XL-1/W2, Rockstock, Germany, 4–6 September 2013; pp. 281–286.

60. Agüera-Vega, F.; Carvajal-Ramírez, F.; Martínez-Carricondo, P. Accuracy of Digital Surface models and Orthophotos Derived from Unmanned Aerial Vehicle Photogrammetry. *J. Surv. Eng.* **2017**, *143*. [CrossRef]

61. Rango, A.; Laliberte, A. Impact of flight regulations on effective use of unmanned aircraft systems for natural resources applications. *J. Appl. Remote Sens.* **2010**, *4*, 043539. [CrossRef]

 *forests*

Article

Automatic Segmentation of *Mauritia flexuosa* in Unmanned Aerial Vehicle (UAV) Imagery Using Deep Learning

Giorgio Morales *, Guillermo Kemper, Grace Sevillano, Daniel Arteaga, Ivan Ortega and Joel Telles

National Institute of Research and Training in Telecommunications (INICTEL-UNI), National University of Engineering, Lima 15021, Peru; guillermo.kemper@gmail.com (G.K.); ksevillano@uni.pe (G.S.); darteaga@inictel-uni.edu.pe (D.A.); ivan.ortega91@gmail.com (I.O.); jtelles@inictel-uni.edu.pe (J.T.)
* Correspondence: gmorales@inictel-uni.edu.pe

Received: 30 October 2018; Accepted: 23 November 2018; Published: 26 November 2018

Abstract: One of the most important ecosystems in the Amazon rainforest is the *Mauritia flexuosa* swamp or "aguajal". However, deforestation of its dominant species, the *Mauritia flexuosa* palm, also known as "aguaje", is a common issue, and conservation is poorly monitored because of the difficult access to these swamps. The contribution of this paper is twofold: the presentation of a dataset called MauFlex, and the proposal of a segmentation and measurement method for areas covered in *Mauritia flexuosa* palms using high-resolution aerial images acquired by UAVs. The method performs a semantic segmentation of *Mauritia flexuosa* using an end-to-end trainable Convolutional Neural Network (CNN) based on the Deeplab v3+ architecture. Images were acquired under different environment and light conditions using three different RGB cameras. The MauFlex dataset was created from these images and it consists of 25,248 image patches of 512×512 pixels and their respective ground truth masks. The results over the test set achieved an accuracy of 98.143%, specificity of 96.599%, and sensitivity of 95.556%. It is shown that our method is able not only to detect full-grown isolated *Mauritia flexuosa* palms, but also young palms or palms partially covered by other types of vegetation.

Keywords: *Mauritia flexuosa*; semantic segmentation; end-to-end learning; convolutional neural network; forest inventory

---

1. Introduction

The *Mauritia flexuosa* L. palm is the main species of one of the most remarkable ecosystems of the Amazon rainforest: the *Mauritia flexuosa* swamp, also known as "aguajal" [1–3]. Its importance is not only ecological but also social and economic. It is the ecosystem with the greatest carbon dioxide absorption capacity in the Amazon [4,5] and it is habitat of a wide range of fauna [1]. In addition, due to high demand of *Mauritia flexuosa* fruit and derivatives, this species is a key economic engine for the indigenous populations and contributes to their economic and social development [3,6]. Unfortunately, in spite of the stringent government efforts to control deforestation, cutting down M. *Flexuosa* palm trees to harvest their fruits is a common activity [1]. For trees that are harvested, the proportion that is cut versus climbed is unknown, which is why carrying out multidisciplinary studies regarding species population assessment and extraction locations would help to target conservation and management efforts in communities that are hot-spots for extraction [7,8].

Recently, there has been a drastic increase in the use of Unmanned Aerial Vehicles (UAVs) for forest applications due to their low cost, automation capabilities, and the fact that they can support different types of payloads, e.g., RGB or multispectral cameras, LiDAR (Light detection and

Ranging), radar, etc. For instance, UAV photogrammetric data is used to rapidly detect tree stumps or coniferous seedlings in replanted forest harvest areas using basic image processing and machine learning techniques [9,10]. Similarly, UAVs have been used to tackle the problem of tree detection from many perspectives. For example, LiDAR-based methods model the 3D-shape of trees for detection with accuracy values ranging from 86% to 98% [11,12]; however, the high cost of LiDAR for UAVs represents an important limitation. The same limitation occurs with hyperspectral-based methods, such as [13], which uses a hyperspectral frame format camera and an RGB camera along with 3D modelling and Multilayer Perceptron (MLP) neural networks, and obtains accuracy values ranging from 40% to 95% depending on the conditions of the area. Following the idea of exploiting the 3D-shape of trees, some methods perform tree detection from RGB images using generated Digital Surface Models (DSMs), Structure-from-Motion (SfM) or local-maxima based algorithms on UAV-derived Canopy Height Models (CHMs) [14,15]. Nevertheless, the aforementioned methods are likely to show poor performance for trees with irregular canopy, trees in mixed-species forests, or trees that are partially occluded by taller trees.

There exist tree detection methods that use multispectral or RGB cameras and specific descriptors such as crown size, crown contour, foliage cover, foliage color and texture [16]; while others rely on pixel-based classification techniques, such as calculating the Normalized Difference Vegetation Index (NDVI), Circular Hough Transform (CHT) and morphological operators to segment palm trees with an accuracy of 95% [17]. Other methods depend on object-based classification techniques; for example, they use the Random Forest algorithm on multispectral data with an accuracy value of 78% [18], or a naive Bayesian network on high-resolution aerial ortophotos and ancillary data (Digital Elevation Models and forest maps) with an accuracy value of 87% [19].

In recent years, the availability of large datasets and optimal computational resources has allowed for the development of different deep learning techniques, which have now become a benchmark for tackling computer vision problems such as object detection or segmentation. Nevertheless, to the best of our knowledge, few deep learning-based techniques have been proposed to solve the problem of tree detection in aerial images. For instance, the method in [20] used the AlexNet CNN (Convolutional Neural Network) architecture with a sliding window for palm tree detection and counting, obtaining an overall accuracy of 95% over QuickBird images with a spatial resolution of 2.4 m. Similarly, the method in [21] used a pre-trained CNN in combination with the YOLOv2 algorithm to detect Cohune palm trees (*Attalea cohune* C.), with an average precision of 79.5%, and deciduous trees, with an average precision of 67.3%. Furthermore, the method in [22] used Google's CNN Inception v3 with transfer learning and sliding windows to detect coconut trees with a precision of 71% and a recall of 93%. Finally, the method in [23] first segmented aerial forest images into individual tree crowns using the eCognition software and then trained the GoogLeNet model to classify seven tree types with an accuracy of 89%. It is worth mentioning that all of these methods are trained to classify visible tree crowns in the images but do not attempt to delineate or segment the tree crowns; as a consequence, if most of a tree crown is covered by taller trees, trained CNNs are not likely to detect it.

In this work, we present a new efficient method to semantically segment *Mauritia flexuosa* palm trees in aerial images acquired with RGB cameras mounted on Unmanned Aerial Vehicles (UAV). Our aerial images of a *Mauritia flexuosa* swamp located south of the Peruvian city of Iquitos were obtained with three different cameras under different climate conditions. By doing so, we created a publicly available dataset of 25,248 image patches of 512 × 512 pixels, each of them with their respective hand-drawn ground truth. With this dataset, we trained five state-of-the-art segmentation deep learning models and decided to use a model based on the Deeplab v3+ architecture [24], as it showed the best performance. The model was trained to detect and segment *Mauritia flexuosa* crowns at different growing stages and scales, even when only a small part of the crown was visible.

2. Materials and Methods

2.1. Mauritia flexuosa

The *Mauritia flexuosa* swamp, also known as "aguajal", is a swamp (humid forest ecosystem) in permanently flooded depressions. Although it is home to more than 500 flora species and 12 fauna species, its dominant species is the *Mauritia flexuosa* palm, also known as "aguaje", which is a palm tree that belongs to the family Arecaceae. In the adult stage, aguajes can grow up to 40 meters (131 feet) in height and 50 centimeters (1.6 feet) in trunk diameter; their leaves are large and form a rounded crown (Figure 1). Each palm tree has an average of eight clusters of fruit, and each cluster produces more than 700 oval-shaped drupes covered in dark red scales [1].

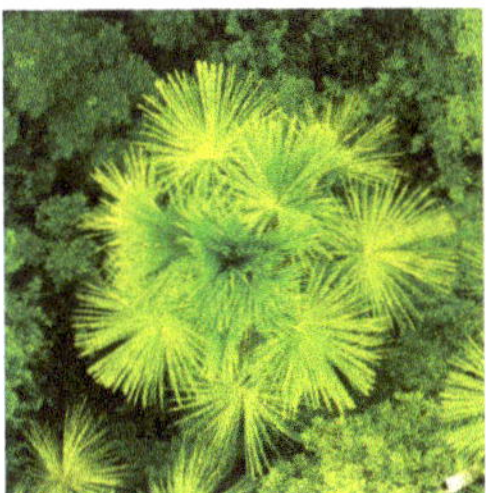

Figure 1. Aerial view of a *Mauritia flexuosa* palm.

The extent of *Mauritia flexuosa* swamps in the Peruvian Amazon rainforest is quite significant. An example is the Ucamara depression between the Ucayali and Marañón rivers, in the region of Loreto, whose capital is the Iquitos City. There, the extent of these swamps reaches about four million hectares (10% of the region surface) [3].

In addition to the economic (Iquitos City alone consumes up to 50 metric tons of aguaje a day) [1], social [3] and nutritional value [25] of this palm tree, its environmental importance is also to be highlighted: in 2010, the FAO Forestry Department stated that, for the evaluation period 2002–2008 in an area of 1,415,100 hectares of aguajales, 146,462,850 metric tons of carbon were stored in vegetation (103.5 t/ha) and 141,510,000 metric tons of carbon in soil (100 t/ha), which represents the greatest carbon absorption capacity of all ecosystems in the Amazonian rainforest [5].

Worryingly, cutting down these trees to harvest the fruit of aguaje is affecting several populations of *Mauritia flexuosa* female palms. It is estimated that 17 million of these palms are cut down in the surroundings of Iquitos to meet the demand of the city [1]. This has resulted in the disappearance of female individuals in accessible *Mauritia flexuosa* populations, thus affecting the food chains of such regions (due to their key importance in the diet of the Amazonian fauna) and causing genetic erosion (since the best and more productive palms are cut down). For such reasons, these ecosystems should be properly and continuously monitored so that preventive measures can be taken in order to prevent illegal logging and the disappearance of this important palm tree.

2.2. Image Acquisition

2.2.1. Study Area

The study area consisted of two regions with different densities of *Mauritia flexuosa*. The one with the higher density was located in the surroundings of Lake Quistococha, south of Iquitos City. The other region was located next to the facilities of the Peruvian Amazon Research Institute (IIAP). Both areas are in Iquitos City, in Maynas Province. Figure 2 shows six orthomosaics corresponding to the regions above.

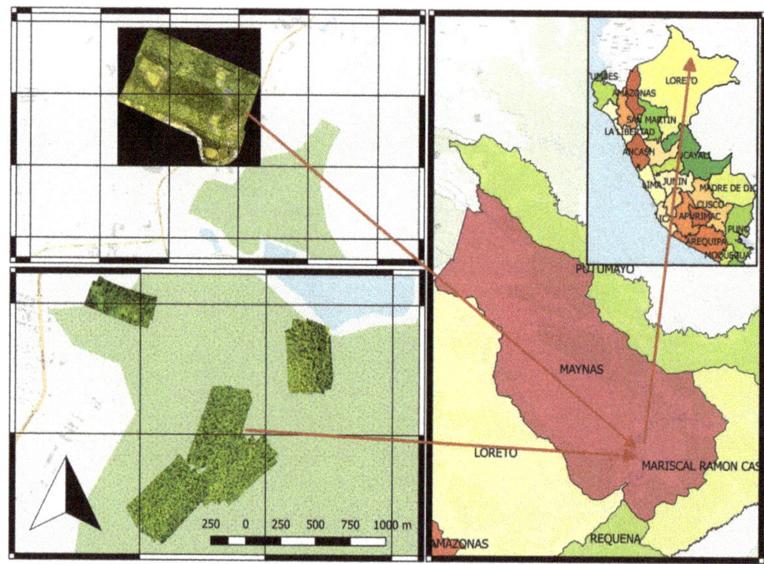

Figure 2. Study area in Iquitos City, Maynas Province, north of Peru.

2.2.2. UAV Imagery

UAV imagery was collected over the years (2015, 2016, 2017 and 2018) under different atmospheric conditions. The flight crew consisted of two pilots and one spotter. We used three UAVs with different camera models; and so, we acquired images with different features. Further details are summarized in Table 1.

Table 1. Unmanned aerial vehicles (UAVs) and cameras specifications.

UAV Specifications			
Description	Quadcopter	Quadcopter	Quadcopter
Brand	Aeryon	DJI	TurboAce
Model	SkyRanger sUAS	Mavic Pro	Matrix-E
Vehicle Dimensions	1020 × 1020 × 240 mm	485 × 430 × 83 mm	1160 × 840 × 250 mm
Vehicle Weight (kg)	2.4	0.734	4
Camera Specifications			
Camera Model	Aeryon MT9F002	DJI FC220	Sony Nex-7
Image Size (megapixels)	14 MP	12 MP	24 MP
Ground Sampling Distance	1.4 cm/pixel	2.5 cm/pixel	1.4 cm/pixel
Flight Altitude	80 m	70 m	100 m
Image Dimensions (pixels)	4608 × 3288	4000 × 3000	4000 × 6000
Bit Depth	24	24	24

The Sony Nex-7 camera mounted in the Matrix-E UAV was manually configured: the ISO value was 200; the maximum aperture was *f/8*; and the shutter speed was 1/320. The settings of the SkyRanger and the Mavic Pro cameras were set to automatic. Many of the images were acquired near midday with cloud-free conditions (Figure 3a); however, Iquitos is normally covered in big clouds, and that is why we obtained some dark images of forest under shadows (Figure 3b). Some images were also acquired in the afternoon, and due to the angle of incidence of the sun's rays, there were many shadows cast by tall trees (Figure 3c). Moreover, the images acquired with the SkyRanger camera showed a defect around the corners known as vignetting (Figure 3d). Finally, because we flew at different altitudes, we achieved Ground Sample Distances (GSD) from 1.4 to 2.5 cm/pixel. In summary, we acquired images with different resolutions, white balance settings, light conditions and others defects; nevertheless, *Mauritia flexuosa* palms can still be recognized by any trained human.

Figure 3. Aerial images acquired by different UAVs. (**a**) Cloud-free region captured with a Sony Nex-7. (**b**) Shadowed region captured with a Sony Nex-7. (**c**) Aerial image acquired in the afternoon with a Sony Nex-7. (**d**) Aerial image captured by the Skyranger UAV with vignetting. (**e**) and (**f**) Aerial images captured by the Mavic Pro UAV.

2.2.3. MauFlex Dataset

Among all the aerial images acquired over the last four years, we selected 96 of the most representative to create the dataset: 47 were acquired by the TurboAce UAV; 28, by the Mavic Pro UAV; and 21, by the SkyRanger UAV. Each image has a binary hand-drawn mask indicating the presence of *Mauritia flexuosa* palms in white. From these images, we extracted image patches of 512 × 512 pixels.

To analyze the images at different scales, the images captured by the TurboAce UAV were resized to 50% and 25% of their original size due to their high level of detail. In addition, we used data augmentation to increase the dataset size and to prevent overfitting issues; thus, each patch was rotated 90°, 180° and 270° [26]. This is how we created the MauFlex dataset (See Supplementary Materials) [27], which is made up of 25,248 image patches, each one with its respective binary mask, as shown in Figure 4. We split 95% of the data to create the training set, 2.5% to create the validation set and 2.5% to create the test set. These three sets are independent among them.

Figure 4. Samples of original images and shadow masks from the MauFlex dataset.

2.3. Proposed CNN for Segmentation

We propose a semantic level segmentation of *Mauritia flexuosa* using a Convolutional Neural Network (CNN). The architecture of our network is based on the Deeplab v3+ architecture [24], which integrates an encoder, a spatial pyramid pooling module, and a decoder. Those modules use inverted residual units, atrous convolutions and atrous separable convolutions, which are briefly described below:

- Inverted residual unit: The main feature of a residual unit is the skip/shortcut between input and output, which allows the network to access earlier activations that were not modified by the convolution blocks, thus preventing network degradation problems such as gradient vanishing or exploding when it is too deep [28]. Inverted residuals units were first introduced in [29]; the main difference is that instead of expanding the number of input channels and then shrinking them, inverted residual units (IRUs) expand the input number of channels using a 1×1 convolution, then apply a 3×3 depthwise convolution (the number of channels remains the same), and, finally, apply another 1×1 convolution that reduces the number of channels, as shown in Figure 5. The IRU shown in Figure 5 uses a batch normalization layer ("BN") and a Rectified-Linear unit layer with a maximum possible value of 6 ("ReLU6") after each convolution layer.

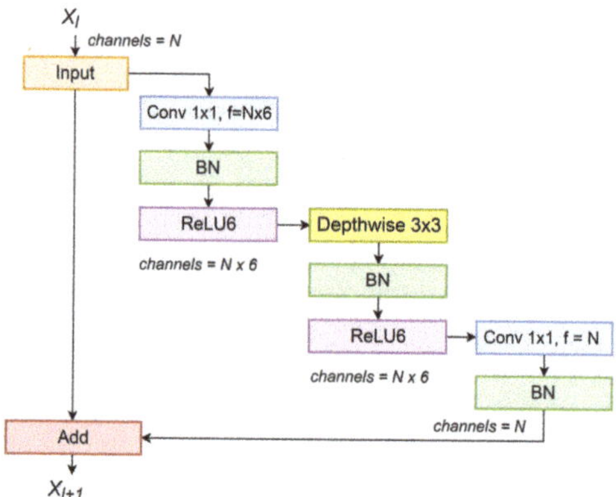

Figure 5. Inverted residual unit (IRU) used in our proposed network. It uses regular 1×1 convolutions ("Conv"), 3×3 depthwise convolutions, batch normalization ("BN") and Rectified Linear Unit activation with a maximum possible value of 6 ("ReLU6").

- Atrous convolution: Also known as dilated convolution, it is basically a convolution with upsampled filters [30]. Its advantage over convolutions with larger filters, is that it allows enlarging the field of view of filters without increasing the number of parameters [31]. Figure 6 shows how a convolution kernel with different dilation rates is applied to a channel. This allows for multi-scale aggregation.

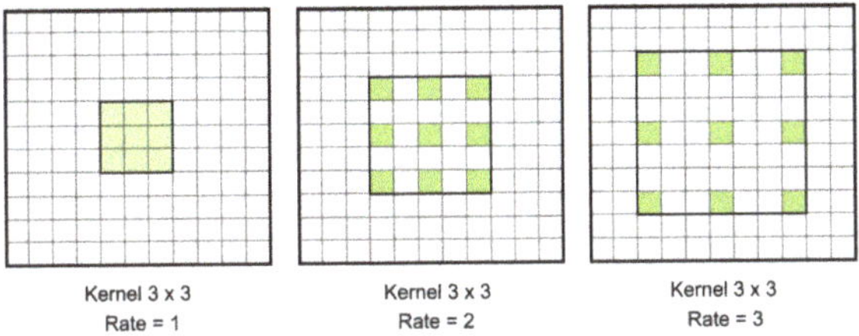

Figure 6. Atrous convolution kernel (green) dilated with different rates.

- Atrous separable convolution: It is a depthwise convolution with atrous convolutions followed by a pointwise convolution [24]. The former performs an independent spatial atrous convolution over each channel of an input; and the latter combines the output of the previous operation using 1×1 convolutions. This arrangement effectively reduces the number of parameters and mathematical operations needed in comparison with a normal convolution.

2.4. CNN Architecture

As we stated before, our proposed architecture is similar to the Deeplab v3+ architecture [24]. Figure 7 shows our architecture and its three main modules: an encoder, an Atrous Spatial Pyramid Pooling (ASPP) module, and a decoder. The main difference from the original Deeplab v3+ network is the number of layers used.

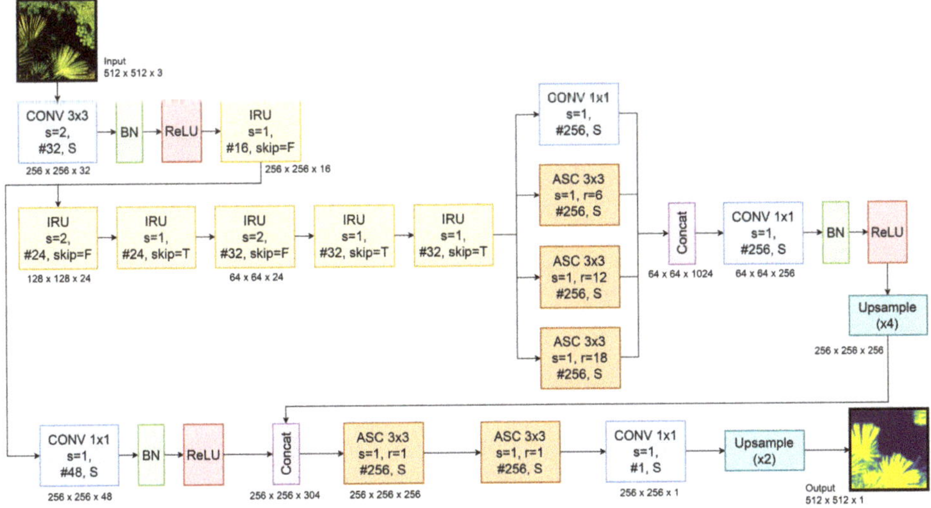

Figure 7. The proposed network architecture. It uses regular convolutions ("CONV"), inverted residual units ("IRU") and atrous separable convolutions ("ASC").

The encoder is a feature extractor that uses several inverted residual units as a backbone and reduces the original size of the image by a factor of eight (*output stride* = 8). The ASPP module applies four parallel atrous separable convolutions with different dilation rates; this allows analyzing the extracted features at different scales. These outputs are concatenated and passed through a 1×1

convolution in order to reduce the number of channels. This result is upsampled by a factor of four and concatenated with low-level features of the same dimension. The motivation for doing so is that the structure in the input should be aligned with the structure in the output, so it is convenient to share information from low levels of the network, such as edges or shapes, to the higher ones. Then, we apply two more 3×3 separable convolutions and finally, a 1×1 convolution with one channel and sigmoid activation, so that a binary mask is obtained. This result is upsampled by a factor of two to recover the original size of the image.

In Figure 7, convolution blocks are denoted as : "CONV;" inverted residual units, as "IRU;" and atrous separable convolution blocks, as "ASC." The output number of filters of each block is reported using the hash symbol ("#"). The stride of all convolutions is denoted as "s." Blocks marked with "S" are "same padded," which means that the output is the same size as the input. "ReLU" represents a standard rectified linear unit activation layer and "BN" a batch normalization layer. If an IRU block is strided, there cannot be a skip between its input and its output; in such cases the "skip" option is set to "False".

3. Results and Discussion

3.1. CNN Training

The training algorithm was implemented using Python 3.6 on a PC with Intel i7-8700 at 3.7 GHz CPU, 64GB RAM and a NVIDIA GeForce GTX 1080 Ti GPU. The proposed CNN was trained using an Adam optimizer [32] with a learning rate of 0.003, a momentum term β_1 of 0.9, a momentum term β_2 of 0.999 and a mini-batch size of 16. The binary cross-entropy function was chosen as our loss function given the fact that it is commonly used for binary segmentation problems and that there is a balance between the amount of pixels of both training classes; thus, it was not necessary to implement specialized loss functions, such as weighted binary cross-entropy function. Figure 8 shows the evolution of network accuracy and loss over training time. After each training epoch, the accuracy and the loss are calculated on the validation set to monitor its ability to generalize and avoid overfitting. The spikes shown in validation loss in epochs 30 and 50, approximately, correspond to a decrease in performance in the training set. This is an expected behaviour during the first training epochs, since the model is still unstable and it is not able to generalize well; however, when the model stabilizes, the validation loss fluctuates with small spikes close to the training loss.

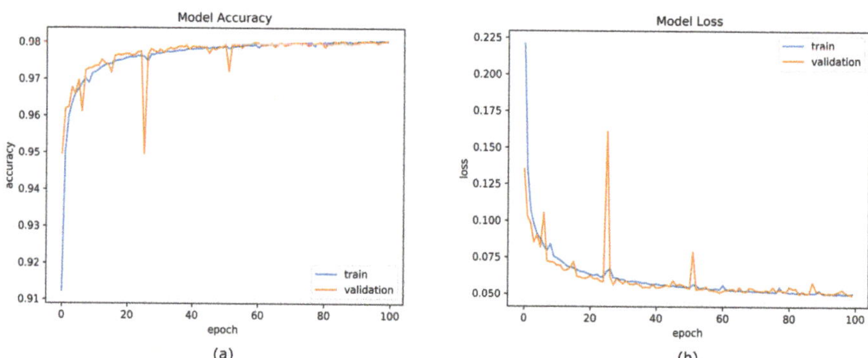

Figure 8. Metrics evolution over training time of our proposed network. (**a**) Epochs vs. Accuracy. (**b**) Epochs vs. Loss.

In order to compare the performance of our proposed network with a different segmentation approach, we trained four other networks based on the U-NET structure [33] to compare the results and choose the best one. A U-NET is a network composed of an encoder and a decoder with skip connections that has been widely used for solving segmentation problems. The encoder-decoder structure of the U-NET tends to extract global features of the inputs and generate new representations from this overall information. Because we experienced a sudden drop in the accuracy metric during training, we decided to strengthen our networks by implementing skips between the input and output of each layer with 1×1 convolutions in order to equalize the number of channels before the addition operation, thus converting our U-NETs to ResU-NETs [34]. The first implemented network (*U-NET1*) has three layers in the encoder and three in the decoder; each layer has a 3×3 convolution block followed by a batch normalization block and a ReLU activation. Furthermore, we added a 10% dropout rate in the decoder layers to prevent overfitting. The second network (*U-NET2*) is similar to the previous one but has four layers in the encoder and four in the decoder. The third (*U-NET3*) and fourth (*U-NET4*) networks have the same structure as the first and the second networks, respectively, but they apply atrous separable convolutions with dilation rates of two instead of regular convolutions. Figure 9 shows the evolution of accuracy and loss of all networks over training time.

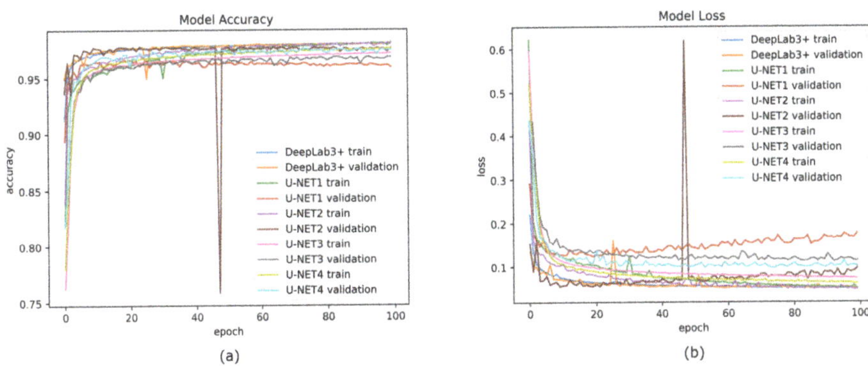

Figure 9. Comparison of metrics evolution over training time of all networks. (**a**) Epochs vs. Accuracy. (**b**) Epochs vs. Loss.

To statistically analyze the behavior of our network against the other networks, we calculated four metrics from the validation set: accuracy (ACC), precision (PREC), recall/sensitivity (SN), and specificity (SP), as shown in Table 2. The ACC ratio indicates correctly predicted observations against total observations; the PREC ratio indicates correctly predicted positive observations against total predicted positive observations; the SN ratio indicates correctly predicted positive observations against total actual positive observations, and the SP ratio indicates correctly predicted negative observations against total actual negative observations. Additionally, the number of trainable parameters of each network is added in Table 2.

Table 2. Metrics Comparison of Different Shadow Detection Methods.

Metric / Method	ACC (%)	PREC (%)	SN (%)	SP (%)	Parameters
U-NET1	95.973	91.381	92.632	97.087	3,736,321
U-NET2	97.682	94.858	95.953	98.261	3,910,641
U-NET3	96.843	92.534	94.886	97.486	503,100
U-NET4	97.512	95.166	95.028	98.358	542,460
Proposed network	98.036	96.688	95.616	98.871	507,729

In Table 2 we observe that our method has achieved the highest metric values. Our method is nearly 0.5% more accurate, sensitive and specific when compared to the second best accuracy, sensitivity and specificity values; and nearly 1.5% more precise when compared to the second best precision value. That means that our proposed network is particularly better than the others are at avoiding false positives. Although these differences may not seem significant, we observe in Figures 8 and 9 that only our method shows a little difference between the training and validation values over the training time, meaning that it prevents overfitting problems and has better performance than the other networks when it comes to predicting new samples outside the training set. Furthermore, we notice a huge difference between the number of trainable parameters of *U-NET1* and *U-NET3*, and *U-NET2* and *U-NET4*, although they have similar architectures, proving that using atrous separable convolutions instead of regular convolutions significantly reduces the amount of computation. Finally, another advantage of our method is that it has 34,731 less parameters than U-NET4; thus, it is faster because it has less operations to perform. When evaluating on the test set, the proposed network showed an accuracy of 98.143%, a specificity of 96.599%, and a sensitivity of 95.556%. This represents an unbiased evaluation of the final selected network.

3.2. *Mauritia flexuosa* Segmentation

Figure 10 shows the segmentation results of 512 × 512 patches; however, one aerial photograph contains several of these small patches, as its dimensions are much larger (Table 1). Thus, to perform the *Mauritia flexuosa* segmentation of a whole image, we apply a 512 × 512 sliding window across the image in both horizontal and vertical direction with a 50-pixel overlap. This sliding window is processed by the trained CNN in each position. Then, the image is reconstructed with the segmentation results, as shown in Figure 11. In order to avoid discontinuities or discrepancies in the overlapping pixels captured by the moving pixels, we always considered the maximum pixel values. Furthermore, a threshold of 0.5 is applied over the probability map (Figure 11b) to obtain a binary mask as shown in Figure 11c.

Figure 10. *Mauritia flexuosa* segmentation results.

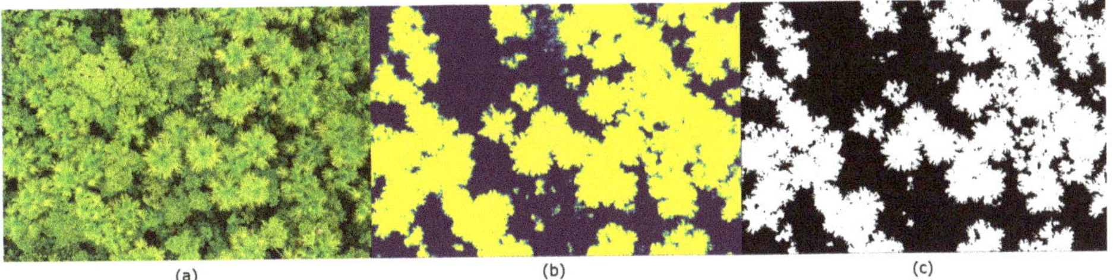

Figure 11. *Mauritia flexuosa* segmentation result for a whole image. (**a**) Original image. (**b**) *Mauritia flexuosa* probability map. (**c**) *Mauritia flexuosa* binary mask.

3.3. Mauritia flexuosa Monitoring

The proposed algorithm is designed to be used as a tool by experts from the Peruvian Amazon Research Institute (IIAP). They will acquire aerial images of areas of interest to monitor periodically the approximate amount of *Mauritia flexuosa* palms on a regular basis.

Hundreds of images can be taken in one single flight; using only one of them is not representative enough to analyze a big area, which is why it is necessary to create a georeferenced image mosaic using the GPS information of each image. The elaboration of a mosaic consists of reconstructing a scene in two dimensions from the combination of images acquired with a certain overlap. To carry out this operation, a series of geometric transformations between pairs of images must be estimated, so that when warping one image on another, they can be blended with the least possible error. For this, we use an algorithm that was specifically developed as part of this project to work on areas with abundant vegetation [35]. Figure 12 illustrates two types of mosaics: one made up of RGB images and the other of binary *Mauritia flexuosa* masks. Figure 13 shows five mosaics of areas with different concentration of *Mauritia flexuosa* palms. By doing this, we can analyze large areas and fly periodically to monitor this kind of natural resources.

Figure 12. Aerial image mosaic composed of 168 photographs. (**a**) Mosaic of RGB images. (**b**) Mosaic of *Mauritia flexuosa* masks.

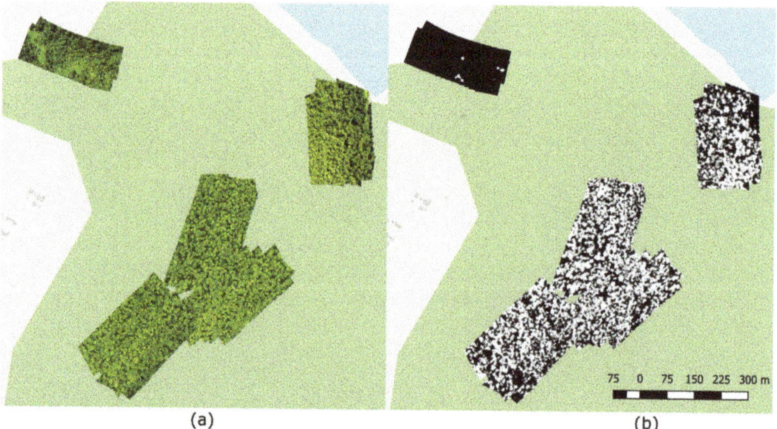

Figure 13. Aerial image mosaics acquired near Lake Quistococha. (**a**) Mosaics of RGB images. (**b**) Mosaics of *Mauritia flexuosa* masks.

4. Conclusions

In this paper, we have presented a new end-to-end trainable deep neural network to tackle the problem of *Mauritia flexuosa* palm trees segmentation in aerial images acquired by Unmanned Aerial Vehicles (UAVs).

The proposed model is based on Google's Deeplab v3+ network and has achieved better performance than those of other Convolutional Neural Networks used for performance comparison. With an accuracy of 98.036%, the segmentation results prove to be quite similar to the hand-drawn ground truth masks. What is more, after learning the particular features of *Mauritia flexuosa* and its leaves (e.g. shape, texture, color, etc.), our model , our model is able to detect the presence of *Mauritia flexuosa* palms and segment them even when partially covered by taller trees. Further work will be focused on both segmenting and counting the approximate amount of *Mauritia flexuosa* palms in high-resolution aerial photographs.

Supplementary Materials: The dataset are available at http://didt.inictel-uni.edu.pe/dataset/MauFlex_Dataset.rar, dataset license: CC-BY-NC-SA 4.0.

Author Contributions: Conceptualization, G.M.; Methodology, G.M.; Software, G.M.; Investigation, G.M., G.S., D.A., and I.O.; UAV Data Acquisition, D.A., I.O., and G.M.; Writing—original draft preparation, G.M.; Writing—review and editing, G.M., G.K. and J.T.; Supervision, G.K.; Project administration, J.T.

Acknowledgments: This research was funded by Programa Nacional de Innovación para la Competitividad y Productividad (Innóvate Perú) grant number 393-PNICP-PIAP-2014. The authors acknowledge the Peruvian Amazon Research Institute (IIAP) for its support during the image acquisition process in the Amazon rainforest.

Conflicts of Interest: The authors declare no conflict of interest. The founding sponsors had no role in the design of the study; in the collection, analyses, or interpretation of data; in the writing of the manuscript, or in the decision to publish the results.

References

1. Del Castillo, D.; Otárola, E.; Freitas, L. *Aguaje: The Amazing Palm Tree of the Amazon*; IIAP: Iquitos, Perú, 2006; ISBN 9972-667-34-0.

2. Freitas, L.; Pinedo, M.; Linares, C.; Del Castillo, D. *Descriptores Para el Aguaje (Mauritia flexuosa L.F.)*; IIAP: Iquitos, Perú, 2006; ISBN 978-9972-667-39-8.

3. Levistre-Ruiz, J.; Ruiz-Murrieta, J. "El Aguajal": El bosque de la vida en la Amazonía peruana. *Cienc. Amaz.* **2011**, *1*, 31–40. [CrossRef]

4. Draper, F.C.; Roucoux, K.H.; Lawson, I.T.; Mitchard, E.T.; Honorio, E.N.; Lähteenoja, O.; Torres, L.; Valderrama, E.; Zaráte, R.; Baker, T.R. The distribution and amount of carbon in the largest peatland complex in Amazonia. *Environ. Res. Lett.* **2014**, *9*, 124017. [CrossRef]

5. Malleux, R.; Dapozzo, B. Evaluación de los recursos forestales mundiales 2010—Informe Nacional Perú. Available online: http://www.fao.org/docrep/013/al598S/al598S.pdf (accessed on 22 October 2018).

6. Mesa, L.; Galeano, G. Palms uses in the Colombian Amazon. *Caldasia* **2013**, *35*, 351–369.

7. Virapongse, A.; Endress, B.A.; Gilmore, M.P.; Horn, C.; Romulo, C. Ecology, livelihoods, and management of the *Mauritia flexuosa* palm in South America. *Glob. Ecol. Conserv.* **2017**, *10*, 70–92. [CrossRef]

8. Ticktin, T. The ecological implications of harvesting non-timber forest products. *J. Appl. Ecol.* **2004**, *41*, 11–21. [CrossRef]

9. Puliti, S.; Talbot, B.; Astrup, R. Tree-Stump detection, segmentation, classification, and measurement using Unmanned Aerial Vehicle (UAV) imagery. *Forests* **2018**, *9*, 102. [CrossRef]

10. Feduck, C.; McDermid, G.J.; Castilla, G. Detection of coniferous seedlings in UAV imagery. *Forests* **2018**, *9*, 432. [CrossRef]

11. Balsi, M.; Esposito, S.; Fallavollita, P.; Nardinocchi, C. Single-tree detection in high-density LiDAR data from UAV-based survey. *Eur. J. Remote Sens.* **2018**, *51*, 679–692. [CrossRef]

12. Wallace, L.; Lucieer, A.; Watson, C.S. Evaluating tree detection and segmentation routines on very high resolution UAV LiDAR data. *IEEE Trans. Geosci. Remote Sens.* **2014**, *52*, 7619–7628. [CrossRef]

13. Nevalainen, O.; Honkavaara, E.; Tuominen, S.; Viljanen, N.; Hakala, T.; Yu, X.; Hyyppä, J.; Saari, H.; Pölönen, I.; Imai, N.N.; et al. Individual tree detection and classification with UAV-Based photogrammetric point clouds and hyperspectral imaging. *Remote Sens.* **2017**, *9*, 185. [CrossRef]

14. Klein, A.M.; Dalla, A.P.; Péllico, S.; Strager, M.P.; Schoeninger, E.R. Treedetection: Automatic tree detection using UAV-based data. *Floresta* **2018**, *48*, 393–402. [CrossRef]

15. Mohan, M.; Silva, C.A.; Klauberg, C.; Jat, P.; Catts, G.; Cardil, A.; Hudak, A.T.; Dia, M. Individual tree detection from unmanned aerial vehicle (UAV) derived canopy height model in an open canopy mixed conifer forest. *Forests* **2017**, *8*, 340. [CrossRef]

16. Trichon, V. Crown typology and the identification of rain forest trees on large-scale aerial photographs. *Plant Ecol.* **2001**, *153*, 301–312. [CrossRef]

17. Al Mansoori, S.; Kunhu, A.; Al Ahmad, H. Automatic palm trees detection from multispectral UAV data using normalized difference vegetation index and circular Hough transform. In Proceedings of the SPIE Remote Sensing Conference 10792, Berlin, Germany, 10–13 September 2018; doi: 10.1117/12.2325732.

18. Franklin, S.E.; Ahmed, O.S. Deciduous tree species classification using object-based analysis and machine learning with unmanned aerial vehicle multispectral data. *Int. J. Remote Sens.* **2018**, *39*, 5236–5245. [CrossRef]

19. Mukashema, A.; Veldkamp, A.; Vrieling, A. Automated high resolution mapping of coffee in Rwanda using an expert Bayesian network. *Int. J. Appl. Earth Obs. Geoinf.* **2014**, *33*, 331–340. [CrossRef]

20. Li, W.; Fu, H.; Yu, L.; Cracknell, A. Deep learning based oil palm tree detection and counting for high-resolution remote sensing images. *Remote Sens.* **2017**, *9*, 22. [CrossRef]

21. Epperson, M. Empowering Conservation through Deep Convolutional Neural Networks and Unmanned Aerial Systems. Master's Thesis, University of California, Oakland, CA, USA, 2018.

22. Zakharova, M. Automated Coconut Tree Detection in Aerial Imagery Using Deep Learning. Master's Thesis, The Katholieke Universiteit Leuven, Löwen, Belgium, 2017.

23. Onishi, M.; Ise, T. Automatic classification of trees using a UAV onboard camera and deep learning. *arXiv* **2018**, arXiv:1804.10390.

24. Chen, L.-C.; Zhu, Y.; Papandreou, G.; Schroff, F.; Adam, H. Encoder-Decoder with Atrous Separable Convolution for Semantic Image Segmentation. In Proceedings of the European Conference on Computer Vision (ECCV 2018), Munich, Germany, 8–14 September 2018.

25. Ministry of Health of Peru, 2009. Tablas Peruanas de Composición de Alimentos. Available online: http://www.ins.gob.pe/insvirtual/images/otrpubs/pdf/Tabla%20de%20Alimentos.pdf (accessed on 22 October 2018).

26. Wang, J.; Perez, L. The Effectiveness of Data Augmentation in Image Classification using Deep Learning. *arXiv* **2018**, arXiv:1712.04621.

27. National Institute of Research and Training in Telecommunications (INICTEL-UNI), 2018. MauFlex Dataset. Available online: http://didt.inictel-uni.edu.pe/dataset/MauFlex_Dataset.rar (accessed on 22 October 2018).

28. He, K.; Zhang, X.; Ren, S.; Sun, J. Deep residual learning for image recognition. In Proceeding of the IEEE Conference on Computer Vision and Pattern Recognition (CVPR), Las Vegas, NV, USA, 27–30 June 2016; pp. 770–778.

29. Sandler, M.; Howard, A; Zhu, M.; Zhmoginov, A.; Chen, L.-C. MobileNetV2: Inverted Residuals and Linear Bottlenecks. *arXiv* **2018**, arXiv:1801.04381.

30. Yu, F.; Koltun, V. Multi-scale context aggregation by dilated convolutions. In Proceedings of the International Conference on Learning Representations (ICLR 2016), San Juan, PR, USA, 2–4 May 2016.

31. Chen, L.-C.; Papandreou, G.; Kokkinos, I.; Murphy, K.; Yuille, A.L. DeepLab: Semantic Image Segmentation with Deep Convolutional Nets, Atrous Convolution, and Fully Connected CRFs. *IEEE Trans. Pattern Anal. Mach. Intell.* **2018**, *40*, 834–848. [CrossRef] [PubMed]

32. Kingma, D.; Ba, J. Adam: A method for stochastic optimization. In Proceedings of the International Conference on Learning Representations (ICLR 2015), San Diego, CA, USA, 7–9 May 2015.

33. Ronneberger, O.; Fischer, P.; Brox, T. U-net: Convolutional networks for biomedical image segmentation. In Proceedings of the Medical Image Computing and Computer-Assisted Intervention (MICCAI 2015), Munich, Germany, 5–9 October 2015; Volume 9351, pp. 234–241.

34. Zhang, Z.; Liu, Q.; Wang, Y. Road Extraction by Deep Residual UNet. *IEEE Geosci. Remote Sens. Lett.* **2018**, *15*, 749–753. [CrossRef]

35. Arteaga, D. Desarrollo de un Aplicativo de Software Basado en Algoritmos de Procesamiento Digital de Imágenes y Visión Computacional, Orientado a la Construcción y Georreferenciación de Mosaicos de Imágenes Aéreas Adquiridas vía UAV. Bachelor's Thesis, Universidad Nacional de Ingeniería, Rímac, Peru, 2018.

 *forests*

Article

Application of UAV Photogrammetric System for Monitoring Ancient Tree Communities in Beijing

Zixuan Qiu [1], Zhong-Ke Feng [1,*], Mingming Wang [1], Zhenru Li [2] and Chao Lu [3]

[1] Precision Forestry Key Laboratory of Beijing, Beijing Forestry University, Beijing 100083, China; baronq@foxmail.com (Z.Q.); chinawmm@126.com (M.W.)

[2] Daxing District Gardening and Greening Bureau, Beijing 102600, China; yljcyk@bjdx.gov.cn

[3] Beijing Daxing Fruit and Forestry Institute, Beijing 102600, China; luchaoxin2011@163.com

* Correspondence: zhongkefeng@bjfu.edu.cn; Tel.: +86-138-1030-5579

Received: 18 October 2018; Accepted: 23 November 2018; Published: 24 November 2018

Abstract: Ancient tree community surveys have great scientific value to the study of biological resources, plant distribution, environmental change, genetic characteristics of species, and historical and cultural heritage. The largest ancient pear tree communities in China, which are rare, are located in the Daxing District of Beijing. However, the environmental conditions are tough, and the distribution is relatively dispersed. Therefore, a low-cost, high-efficiency, and high-precision measuring system is urgently needed to complete the survey of ancient tree communities. By unmanned aerial vehicle (UAV) photogrammetric program research, ancient tree information extraction method research, and ancient tree diameter at breast height (DBH) and age prediction model research, the proposed method can realize the measurement of tree height, crown width, and prediction of DBH and tree age with low cost, high efficiency, and high precision. Through experiments and analysis, the root mean square error (RMSE) of the tree height measurement was 0.1814 m, the RMSE of the crown width measurement was 0.3292 m, the RMSE of the DBH prediction was 3.0039 cm, and the RMSE of the tree age prediction was 4.3753 years, which could meet the needs of ancient tree survey of the Daxing District Gardening and Greening Bureau. Therefore, a UAV photogrammetric measurement system proved to be capable when applied in the survey of ancient tree communities and even in partial forest inventories.

Keywords: UAV photogrammetry; forest modeling; ancient trees measurement; tree age prediction

1. Introduction

Ancient trees are the cornerstone of the natural, agricultural, and urban ecosystems on earth [1–5]. The investigation of ancient tree communities is of great scientific value to the study of biological resources, plant distribution, environmental change, genetic characteristics of species, and historical and cultural heritage [6,7]. In view of the specific situation of the ancient tree community survey, few unique survey patterns have been formed. At present, the traditional forest survey pattern is mainly used in the survey of ancient tree communities. The Daxing District of Beijing has the world's rarest ancient pear tree communities. For the purpose of protecting and managing ancient trees, preventing the malicious and illegal felling of ancient trees, and strengthening the information management of ancient trees, the Daxing District Gardening and Greening Bureau hopes to carry out ancient tree communities monitoring every year. As more than 100,000 pear trees and 40,000 ancient pear trees are scattered over more than 400 km^2, great difficulties have arisen in the investigation of ancient tree communities. In addition, the Daxing District Gardening and Greening Bureau can only conduct the survey of ancient pear trees within one month before the Pear Flower Festival every year. Therefore, the high efficiency and real-time performance of the survey of ancient tree communities becomes particularly critical.

Traditional forest survey patterns include ground surveys and aerial remote sensing surveys [8,9]. As ground survey equipment, an electronic total station can accurately measure the three-dimensional coordinates, height, diameter at breast height (DBH), stem volume, and canopy volume of an individual tree [10]. Forest intelligent surveying and mapping instruments, which are portable and internal-external integrated, can be applied in the measurement of height, DBH, and volume of an individual tree. In addition, stand average height, stand average DBH, stand density, and stand volume can be estimated by the angle gauge measure function embedded in the program [11]. A forest telescope intelligent dendrometer is capable of measuring height and DBH of an individual tree from a long range. In addition, stand parameters can be estimated by using the embedded micro sample plot measure function [12]. Even though the above survey instruments have the function of stand parameter estimation, they are more suitable for the accurate measurement of an individual tree than a large-scale forest survey. A terrestrial laser scanning (TLS) system, as an efficient and high-precision measurement method, has gradually become an important means of conducting a forest survey and can visually measure stand structure parameters in three dimensions [13,14]. The main purpose of TLS is to improve the efficiency of forest sample plot monitoring. With the help of the point cloud, automatic acquisition replaces manual measurement of tree attributes, which mainly include tree height, DBH, crown width, and coordinates [15]. The line of sight of TLS is not limited to only a few meters. Several scanning locations are required rather to avoid gaps in the point cloud due to occlusion from terrain or vegetation [16]. This limitation means that using TLS point cloud data to describe large areas of forest space is time-consuming and costly [17].

Airborne laser scanning (ALS) is a good solution for large areas of forest investigation. The ALS systems can generate 3D point cloud data to describe tree height and canopy structure and use other methods to build the relationship between tree heights, canopy structures, and other forest attributes [18]. Many research results showed that the accuracy and precision of the forest survey were satisfying in obtaining forest attributes, such as forest volume and forest biomass, using ALS systems [19–23]. More and more people choose to use high-resolution digital aerial images to generate 3D data just like ALS and apply it to the forest survey [24–26]. The cost of unmanned aerial vehicle (UAV) photogrammetric measurement systems is more acceptable compared to ALS, and attributes such as species composition, maturity, and health status can be acquired through images. Many studies have shown that UAV photogrammetric measurement systems can generate 3D data and the combination of the digital surface model (DSM) and digital elevation model (DEM) can deduce the height of a canopy on the ground, thus producing a canopy height model (CHM) [27]. Generally, UAV photogrammetric measurement systems rely on a ground control point (GCP), which is regarded as a source of reliable georeferencing information [28]. Furthermore, in the UAV photogrammetric measurement systems, additional GCPs are often used to calibrate the location parameters [28]. In general, UAV photogrammetric measurement systems require more than 30 control points per square kilometer, which is undoubtedly extremely difficult and inefficient in a complex forest environment [29]. Therefore, we need a kind of UAV photogrammetric measurement system that can meet the forestry survey requirements. In addition, without a proper sensor geometric calibration, the GCPs will not provide enough accuracy for coordinate correction and that will affect the final 3D model accuracy. Nan An and others imported the converted TIFF images into the Agisoft Photoscan program to generate an orthophoto by correcting perspective distortion [30]. Dongwook Kim and others used Pix4Dmapper software to auto-compensate the principal point and radial distortion by processing the bundle block adjustment [31].

Forest modeling is a type of ecological modeling. With the development of forestry informatization, the relationship between forest modeling and high spatial resolution three-dimensional (3D) remote sensing has become closer [32–35]. Temuulen T. Sankey and others have proposed combining light detection and ranging (LiDAR) data and hyperspectral data for ecological modeling and subtle environmental change detection [36,37]. Remote sensing techniques, in combination with forest modeling, which can be applied to estimate forest biomass and carbon stock [38,39] and to monitor forest harvest and recruitment [40], are widely used in ecosystem process modeling [41]. Remote sensing techniques such as LiDAR can be combined with the algorithm for individual tree detection [42–44] to obtain tree height and canopy area. Many studies have shown that the use of linear mixed models can well establish the correlation between tree height, canopy area, and DBH [45]. In addition, in the forest modeling study, if DBH and tree height are known, tree growth modeling combined with forest environment can predict tree age well [46,47]. However, the development of these two models requires an expensive monitoring system to effectively monitor the forest environment [48]. The survey of ancient tree communities in Daxing District requires high efficiency and low cost. Therefore, a new forest modeling method is needed to predict DBH and tree age.

This study aims to solve the following main problems:

1. On the premise of ensuring the accuracy of forestry survey, it is necessary to study a continuous photogrammetric algorithm and software suitable for monitoring ancient tree communities.
2. Using existing technology and algorithms, it is necessary to study the effective and accurate extraction of forest information from the point cloud data.
3. Because of the specific situation of ancient tree communities, it is necessary to study the ancient tree structure relationship model and ancient tree growth model to estimate the DBH and age.

2. Materials and Methods

2.1. Profile of Study Sites

Daxing District (Figure 1) is located in the south area of Beijing, with Tongzhou District to the east, Gu'an County and Bazhou City in the Hebei province in the south, the Yongding River in the west, and the Fangshan, Fengtai, and Chaoyang districts in the north. It has an east longitude of 116°13′–116°43′ and a north latitude of 39°26′–39°51′. The whole district is on the Yongding River alluvial plain. The terrain gradually slopes from the west to the southeast, with an altitude between 14 and 52 m. There are six main rivers in the Daxing District, including the Yongding, Liangshui, Tiantang, Dalong, Xiaolong, and Xinfeng. The Beijing Daxing Wanmu pear orchard is the largest ancient pear tree community with the largest planting area, the earliest flowering, and the most varieties around Beijing. The central area is located in Lihua Village, where more than 40,000 ancient pear trees (Figure 2) have been preserved for over 50 years. There are more than 40 varieties of pear trees in this area, one of which is 417 years old and has been named "gold yellow" by an emperor of Qing dynasty. From Appendix A, we can see the distribution of ancient trees in each town in Daxing District.

2.2. Technical Information

In this study, by using the YS-500 Fixed-Wing UAV (Figure 3 and Table 1, Beijing Global Forest Technology Co. LTD, Beijing, China) and Sony-A7R camera (Table 2), aerial images of Panggezhuang Town were obtained. The flight area was about 49 km^2, and the ground resolution was better than 5 cm. A total of 4984 images were taken through taking off and landing six times, with an average height of 331 m. The fore-and-aft overlap of route planning was 65%, and the side overlap was 75%.

Control points were measured by Yinhe I RTK (Real Time Kinematic) measurement system (made in China South Surveying & Mapping Instrument Company). The specific parameters of RTK are shown in Table 3.

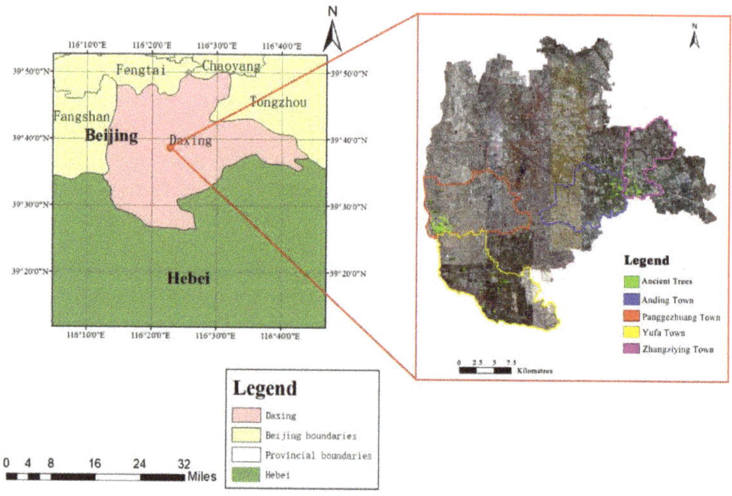

Figure 1. Layout of the research area.

Figure 2. The scene of ancient trees investigation.

Table 1. Parameters of YS-500 Fixed-Wing UAV.

Content	Parameters Value
Wingspan	1700 mm
Length	900 mm
Dynamical System	Dynamoelectric system
Materials	Aeronautical Composite
Maximum Take-off Weight	5.5 kg
Aircraft Standard Load	1.5 kg
Speed	70–120 km/h
Cruising Speed	80–90 km/h
Maximum Flying Distance	100 km
Cruising Range	80 km
Cruising Duration	90 min
Maximum Flying Altitude	4500 m
Wind Resistance	10.8–13.8 m/s
Working Temperature	$-10\,^{\circ}\text{C}$–$50\,^{\circ}\text{C}$
Radio Range	30 km

Table 2. Calibration parameters of the Sony-A7R camera (measurement by the Chinese Academy of Surveying and Mapping).

Calibration Content	Calibration Value
Principal Point x_0	-0.188561
Principal Point y_0	-0.138128
Focal Length f	36.289444 mm
Pixel Size	4.88 μm
Picture Format (pixel)	7360 × 4912
Coefficient of Radial Distortion k_1	-5.145797×10^{-5}
Coefficient of Radial Distortion k_2	1.583367×10^{-7}
Coefficient of Radial Distortion k_3	1.829755×10^{-5}
Tangential Distortion Factor p_1	-6.050453×10^{-6}
Tangential Distortion Factor p_2	$-1.739164\,e^{-5}$

Figure 3. YS-500 Fixed-Wing Unmanned Aerial Vehicle (UAV).

Table 3. Specific parameters of Yinhe I RTK (Real Time Kinematic) measurement system (measurement by the China South Surveying & Mapping Instrument Company).

Instrument Specifications	Specific Parameters
Signal Tracking	220 signal channel BDS B1, B2, B3 GPS L1C/A, L1C, L2C, L2E, L5 GLONASS L1C/A, L1P, L2C/A, L2P, L3 SBAS L1C/A, L5 Galileo GIOVE-A, and GIOVE-B, E1, E5A, E5B QZSS, WAAS, MSAS, EGNOS, GAGAN (Veripos)
GNSS Features	Positioning output frequency, 1 Hz–50 Hz Initialization time, Less than 10 s Able to support GNSS constellation signals from all current and future plans High reliable carrier tracking technology Intelligent dynamic sensitivity positioning technology High precision positioning processing engine
Differential positioning accuracy	Horizontal: 0.25 m + 1 ppm RMS Vertical: 0.50 m + 1 ppm RMS SBAS differential positioning accuracy: Typical <5m 3DRMS
Static GNSS measurements	$\pm(2.5$ mm + 1 mm/km × d), d is the distance of the measured point, km
Real-time dynamic measurement (RTK)	$\pm(8$ mm + 1 mm/km × d), d is the distance of the measured point, km

2.3. Research on the Improved UAV Photogrammetric Program

SfM is the abbreviation of structure from motion, which is a valuable tool for generating 3D models from 2D images. It is developed from computer vision and conventional photogrammetry. Unlike conventional photogrammetry, SfM uses algorithms to identify matching features in the set of overlapping images and to calculate the camera position and direction in accordance with the difference of the multiple matching features [49,50]. Based on these calculations, the overlapping images can be used to reconstruct the "sparse" or "rough" three-dimensional point cloud model of the captured object. The model obtained by the SfM method can be further refined by a multi-view stereo (MVS) algorithm, so as to complete the workflow of SfM-MVS [51].

The SfM-MVS method is relatively inexpensive, both in terms of hardware and software requirements. It is faster than other digital measurements in the field and is a process almost independent of spatial scales. In addition, the SfM-MVS can also produce 3D point cloud data with high precision, high density, and high resolution. In some cases, it may even catch up with the terrestrial laser scanner [49].

In this study, the SfM-MVS method was used to detect the feature points of the image, and the scale-invariant feature transform (SIFT) algorithm (Figure 4) was used to detect feature points and generate feature vectors. By identifying the correspondence between feature points on different images, we screen out the image pairs with overlapping parts. Then, the matching was carried out according to the feature vectors, and the RANSAC (random sample consensus) algorithm was used to delete the connection of the conflicting geometric features of corresponding feature points.

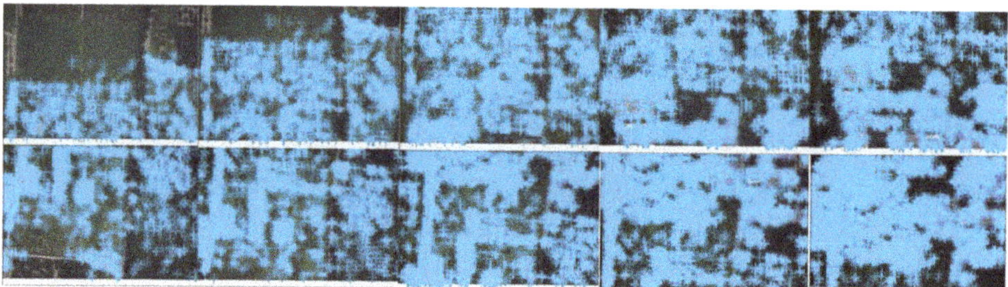

Figure 4. SIFT algorithm image matching the partial region (blue "+" represents feature points that match between images).

The regional network bundle adjustment method of aerial triangulation contains a beam of light composed of an image as the basic unit of adjustment and the collinearity formula of central projection as the basic formula of adjustment [52,53]. Through the rotation and translation of every light beam in space, the best intersection of the light from the common points between the models can be achieved, and the entire region can be optimally incorporated into the known control point coordinate system [54,55]. The advantage of this algorithm is that it only takes more than four GCPs for the free network bundle adjustment method and the system bundle adjustment method, so as to obtain the correction number of the images, improve the efficiency of the field work, and minimize the measurement error.

φ is longitudinal tilt, ω is lateral tilt, and κ is swing angle. X, Y, Z are longitude, latitude, and altitude. Given the profile of the test area, UAV POS (Positioning and Orientation System) data (X_i^0, Y_i^0, Z_i^0, φ_i^0, ω_i^0, κ_i^0) and ground control point $P_n(X_{n0}, Y_{n0}, Z_{n0})$ ($n \geq 4$), free network adjustment was first conducted (Figure 5). Nine control points were arranged in the four corners of the rectangle, the center of the four edges and the center of the rectangle. Among them, ($u_j{'}$, $v_j{'}$) and (u_j, v_j) are the corresponding image point to image $i + 1$ and image i, respectively. λ_j' and λ_j are the corresponding scale factor to image $i + 1$ and image i, respectively. X_{i+1}^0, Y_{i+1}^0, Z_{i+1}^0, φ_{i+1}^0, ω_{i+1}^0, and κ_{i+1}^0 are the elements of exterior orientation of the POS data in the image. $i + 1$, X_i^0, Y_i^0, Z_i^0, φ_i^0, ω_i^0, and κ_i^0 are the elements of exterior orientation of POS data in image i. R_i^0 is the initial value of the rotation matrix composed of the elements of exterior orientation of POS data in image i. R_{i+1}^0 is the initial value of the rotation matrix composed of the elements of exterior orientation of POS data in image $i + 1$. Formula (1) can be obtained by substituting the values of three corresponding image points.

$$
\begin{pmatrix}
1 & 0 & 0 & \lambda_1^{0'}f & 0 & -\lambda_1^{0'}v_1' & a_1^{0'}u_1'+a_2^{0'}v_1'-a_3^{0'}f & 0 & 0 \\
0 & 1 & 0 & 0 & \lambda_1^{0'}f & \lambda_1^{0'}u_1' & b_1^{0'}u_1'+b_2^{0'}v_1'-b_3^{0'}f & 0 & 0 \\
0 & 0 & 1 & \lambda_1^{0'}u_1' & \lambda_1^{0'}v_1' & 0 & c_1^{0'}u_1'+c_2^{0'}v_1'-c_3^{0'}f & 0 & 0 \\
1 & 0 & 0 & \lambda_1^{0'}f & 0 & -\lambda_2^{0'}v_2' & 0 & a_1^{0'}u_2'+a_2^{0'}v_2'-a_3^{0'}f & 0 \\
0 & 1 & 0 & 0 & \lambda_2^{0'}f & \lambda_2^{0'}u_2' & 0 & b_1^{0'}u_2'+b_2^{0'}v_2'-b_3^{0'}f & 0 \\
0 & 0 & 1 & \lambda_2^{0'}u_2' & \lambda_2^{0'}v_2' & 0 & 0 & c_1^{0'}u_2'+c_2^{0'}v_2'-c_3^{0'}f & 0 \\
1 & 0 & 0 & \lambda_3^{0'}f & 0 & -\lambda_3^{0'}v_3' & 0 & 0 & a_1^{0'}u_3'+a_2^{0'}v_3'-a_3^{0'}f \\
0 & 1 & 0 & 0 & \lambda_3^{0'}f & \lambda_3^{0'}u_3' & 0 & 0 & b_1^{0'}u_3'+b_2^{0'}v_3'-b_3^{0'}f \\
0 & 0 & 1 & \lambda_3^{0'}u_3' & \lambda_3^{0'}v_3' & 0 & 0 & 0 & c_1^{0'}u_3'+c_2^{0'}v_3'-c_3^{0'}f
\end{pmatrix}
\begin{pmatrix}
\Delta X^0_{i+1,i} \\ \Delta Y^0_{i+1,i} \\ \Delta Z^0_{i+1,i} \\ \Delta\varphi_{i+1} \\ \Delta\omega_{i+1} \\ \Delta\kappa_{i+1} \\ \Delta\lambda_1' \\ \Delta\lambda_2' \\ \Delta\lambda_3'
\end{pmatrix}
=
\begin{pmatrix}
b_1-\lambda_1^{0'}u_1'+\lambda_1^{0'}\left(a_1^{0'}u_1'+a_2^{0'}v_1'-a_3^{0'}f\right) \\
b_2-\lambda_1^{0'}v_1'+\lambda_1^{0'}\left(b_1^{0'}u_1'+b_2^{0'}v_1'-b_3^{0'}f\right) \\
b_3+\lambda_1^{0'}f+\lambda_1^{0'}\left(c_1^{0'}u_1'+c_2^{0'}v_1'-c_3^{0'}f\right) \\
b_1-\lambda_2^{0'}u_2'+\lambda_2^{0'}\left(a_1^{0'}u_2'+a_2^{0'}v_2'-a_3^{0'}f\right) \\
b_2-\lambda_2^{0'}v_2'+\lambda_2^{0'}\left(b_1^{0'}u_2'+b_2^{0'}v_2'-b_3^{0'}f\right) \\
b_3+\lambda_2^{0'}f+\lambda_2^{0'}\left(c_1^{0'}u_2'+c_2^{0'}v_2'-c_3^{0'}f\right) \\
b_1-\lambda_3^{0'}u_3'+\lambda_3^{0'}\left(a_1^{0'}u_3'+a_2^{0'}v_3'-a_3^{0'}f\right) \\
b_2-\lambda_3^{0'}v_3'+\lambda_3^{0'}\left(b_1^{0'}u_3'+b_2^{0'}v_3'-b_3^{0'}f\right) \\
b_3+\lambda_3^{0'}f+\lambda_3^{0'}\left(c_1^{0'}u_3'+c_2^{0'}v_3'-c_3^{0'}f\right)
\end{pmatrix}. \tag{1}
$$

j stands for corresponding image point. λ_j^0 and $\lambda_j^{0'}$, and b_1, b_2, and b_3 are calculated beforehand. u_j, v_j, u_j', v_j', and f are coordinates of the corresponding image point. Therefore, we can get the corrections, such as $\Delta X^0_{i+1,i}$, $\Delta Y^0_{i+1,i}$, $\Delta Z^0_{i+1,i}$, $\Delta\varphi_{i+1}$, $\Delta\omega_{i+1}$, and $\Delta\kappa_{i+1}$.

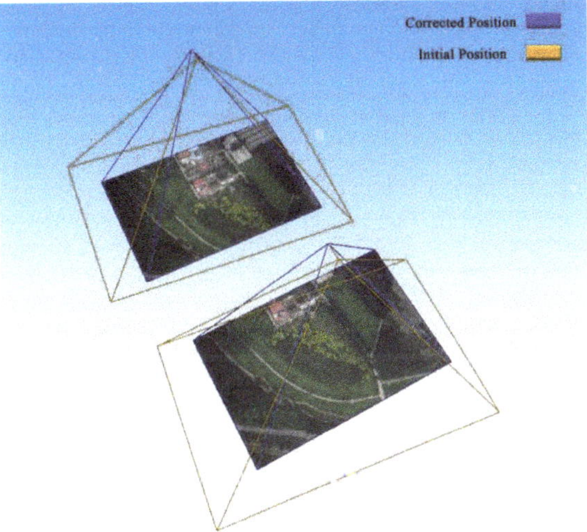

Figure 5. Image correction of the free network adjustment in ancient tree communities.

The selection and positioning of the image control points (Figure 6) is to accurately indicate the position of the image control point on the image. It is the basis of image interpretation and measurement. In the selection and positioning of image control points, the intersection of linear ground objects and the corner of ground objects are generally selected.

According to the correction between images (Figure 7), the adjustment value of each image center is calculated, and the coordinates of each ground control point in the independent coordinate system of images are then calculated. We convert the image coordinate system to the ground coordinate system, so as to obtain the final correction value after the system bundle adjustment method.

From the above, based on the development platform of Microsoft Visual Studio 2010, using C# language, we independently developed software called "New UAV Bundle Adjustment" for image matching and POS correction. The image and POS data processed by this software were imported into Pix4Dmapper software for three-dimensional (3D) points cloud modeling. Pix4Dmapper comes from the Swiss company Pix4D, which is the research result of the world-class research institute EPFL (Swiss Federal Institute of Technology in Lausanne).

Figure 6. Selection and positioning of the image control points.

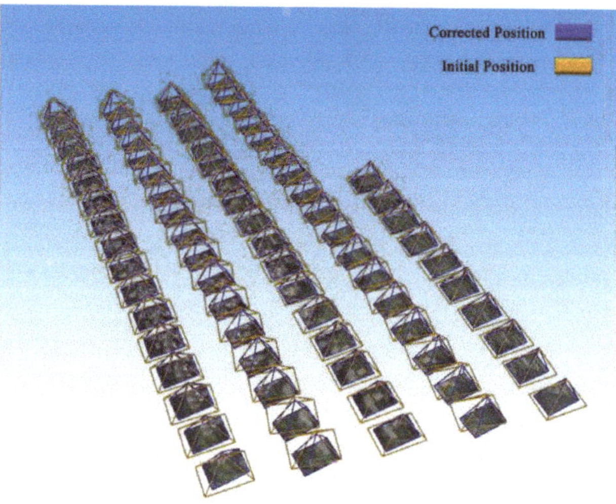

Figure 7. Image correction of the regional system adjustment of ancient tree communities.

The number of control points and the layout position will have a great influence on the precision of aerial triangulation. To verify the accuracy of improved UAV photogrammetric program in this study, nine control points were set up in Panggezhuang Town (Figure A1), and a total of 68 field control points were collected by RTK. The flight area was about 49 km². The same height points were set up on the four sides and the center line of the region. Considering that this test area was a more regular rectangular region, nine points were arranged in the four corners of the rectangle, the center of the four edges, and the center of the rectangle. The remaining 59 points were precision check points. Data acquisition was based on the same camera (Table 2). Data Source A used the improved bundle adjustment for calculation, and Data Source B adopted the conventional aerial measurement method. The layout scheme is shown in Figure 8.

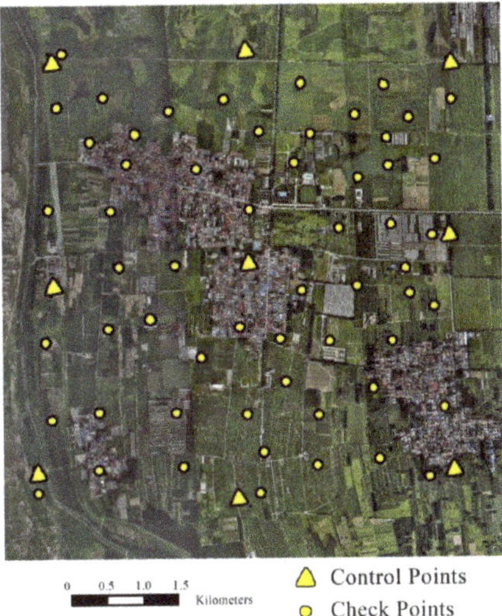

Figure 8. Layout scheme.

The precision of orientation points and check points by using the bundle adjustment was evaluated via field measurement. The difference value between the calculated value of ground coordinate and the measured coordinate was considered to be the true error. The root mean square error was calculated according to Formula (2).

$$\begin{aligned}
\sigma_X &= \sqrt{\frac{\sum(X_C - X_R)^2}{n}} \\
\sigma_Y &= \sqrt{\frac{\sum(Y_C - Y_R)^2}{n}} \\
\sigma_{XY} &= \sqrt{\sigma_X^2 + \sigma_Y^2} \\
\sigma_Z &= \sqrt{\frac{\sum(Z_C - Z_R)^2}{n}}
\end{aligned} \tag{2}$$

In the formula, σ_X and σ_Y were the root mean square error (RMSE) of points in the X and Y directions. σ_Z was the root mean square error of the points in the elevation. σ_{XY} was the root mean square error of the points in the plane. X_C, Y_C, and Z_C were the measured coordinate values of the check points. X_R, Y_R, and Z_R were the calculated coordinate values of check points by using the bundle adjustment.

2.4. Research on the Ancient Tree Information Extraction Method

The image matching point cloud based on the UAV platform can cover large areas and generate high-density accurate point cloud, and the cost is relatively low. An image matching point cloud is a series of inhomogeneous and discrete point sets in space, which contains certain texture information. A DSM can be obtained through the processing of point cloud data, and a DEM can be obtained through the filtering process [49]. As the image point cloud is a passive remote sensing product, it does not have multiple echoes, and it is difficult to obtain accurate DEM data directly in densely vegetated areas. Therefore, it is necessary to conduct research and analysis in information extraction of forest. This study used LiDAR 360 (Beijing Digital Green Earth Technology Co. LTD, Beijing, China) software to extract ancient tree information.

2.4.1. Classification of Ground Point Based on Point Cloud Data

(1) Point cloud denoising. In the process of point cloud acquisition, some noise points will appear due to equipment inaccuracy and environmental factors. Removal of noise points before data

processing can improve data accuracy and reduce the error caused by noise points [56]. The noise removal algorithm adopted in this study includes the high threshold method, the isolated point search method, and the low point search method.

(2) Ground point classification. Ground point refers to the point below ground vegetation or the ground building. The basic idea of extracting the ground point is to assume that the lowest point in an area is its ground point and search these local lowest points to form the initial surface. On this basis, the relationship between other points and the initial surface can be valued. If it conforms to a certain relationship, it will be regarded as the ground point for classification, which is dealt with by multiple iterations. Different geomorphic features have different iterative algorithms and given thresholds [49]. In this study, a hierarchical robust linear predictive filtering algorithm, a morphological filtering algorithm based on gradient, and a TIN stepwise encryption algorithm were used.

2.4.2. Generation of Raster Data from Point Clouds

(1) DSM generation. Point cloud data are irregular three-dimensional discrete points, which need to be interpolated to generate a three-dimensional model with continuous changes [49]. The algorithms used in this study include inverse distance weighted (IDW) interpolation, Kriging interpolation, natural neighbor interpolation, and radial basis function interpolation.

(2) DEM generation. Discrete ground points are interpolated to generate DEM [49]. Firstly, the ground point cloud is rasterized, the ground point is then extracted by the local minimum search window algorithm, and the raster data is interpolated using the TIN interpolation algorithm.

(3) CHM generation. The canopy height model (CHM) is a high-resolution raster dataset that maps the height of the tree to a continuous surface, where each pixel represents the height of the tree above the ground. The method of obtaining CHM is the subtraction of DSM and DEM [49]. Ground fluctuation in the study of ancient tree community causes the bottom of the trees to not be on the same horizontal surface, and the introduction of CHM can solve this problem well. CHM reduces the calculation of tree height to a plane, which can conveniently reflect the information of tree height, as shown in Figure 9.

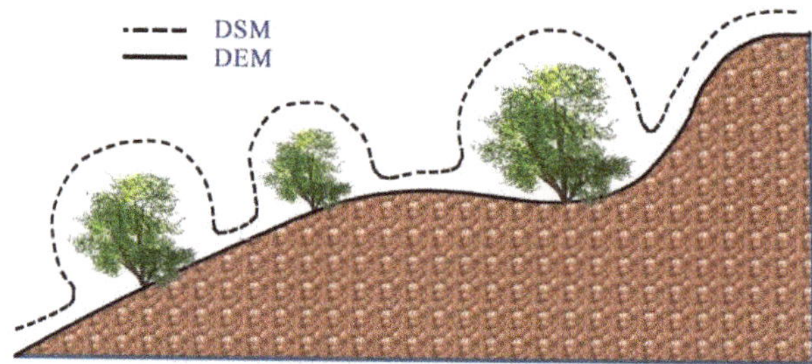

Figure 9. Digital surface model (DSM) and digital elevation model (DEM) schematic.

2.4.3. Segmentation of Individual Ancient Tree

The three-dimensional point cloud is rasterized and transformed into CHM based on the 3D point cloud, from which information such as tree height and crown width can be obtained. In the individual tree segmentation of pear trees, the model algorithm of seed region growth is adopted [49]. The basic method is to form convex hull polygons based on CHM and then reconstruct the canopy along the two-dimensional convex hull with the normalized image point cloud, so as to achieve the purpose of individual tree segmentation. The detailed process is as follows.

(1) A sliding window to detect the location of seed point was defined. The minimum threshold of tree height will be set during processing. When the value is greater than the minimum threshold detected in slide detection, this point is considered as the seed point. (2) The points on CHM are marked and divided into seed points and non-seed points. (3) Four adjacent points near a seed point

are searched for to determine whether the seed point to each point is greater than the set crown width threshold on the plane and whether the height is greater than the height threshold. If the above two points are satisfied, it will be regarded as a new seed point, which will be reclassified and processed several times until all seed points and non-seed points are classified. (4) Marked seed points were used as the center to establish a two-dimensional convex envelope marking boundary. (5) The final boundary contour of the generated point boundary polygon is used as the basis to segment the normalized point cloud data and achieve the purpose of individual tree segmentation.

According to image data and impression of the field situation, the position of missed, excessive, and false detection points were edited. The false seed points could be deleted. The excessive seed points could be deleted and the corrected seed points could be reserved. The missed seed points could be added. Then, the individual tree could be segmented. The central location of the tree could be obtained by the center coordinate of prominent position.

2.5. Research on the Ancient Tree DBH and Age Prediction Model

The height–crown-width–DBH model and "3 speed 2 inflection points" optimum condition growth model of pear trees were developed based on the measured data from a pear tree survey in Lihua Village, Daxing District. In combination with a UAV photogrammetric tree measurement system, the DBH and age of individual trees can be estimated accurately.

2.5.1. Building of the Height–Crown-Width–DBH Model

Data of fixed plots are usually affected by the within-plot and temporal correlation. In order to solve this potential problem of autocorrelation, linear mixed models were fitted to the data by including a random effect of plot (to model spatial correlation) in the model and by specifying an autoregressive error structure (to model temporal correlation) [45]. The fixed plot data of this study is the previous survey data of pear tree communities in Lihua Village (Appendix A, Figure A1), with a total of 1484 pear trees, including only DBH, tree height, and crown width. In 2014, Local Forestry Station investigators made use of the Diameter Ruler and Electronic Total Station (NTS 362R, China South Surveying & Mapping Instrument Company, Guangdong, China) to measure the DBH, height, and crown breadth of pear trees. Therefore, it is more suitable to use the nonlinear model. Referring to the common height–DBH model and crown width–DBH model, the height–crown-width–DBH model (Formula (3)) is obtained.

$$d_{1.3} = g_1 \cdot H^{q_1} + g_2 \cdot D^{q_2}. \tag{3}$$

$d_{1.3}$ is the diameter at breast height, H is the tree height, and D is the crown width. g_1 and q_1 are the factors to establish the correlation between tree height and DBH. g_2 and q_2 are the factors to establish the correlation between crown width and DBH.

2.5.2. Building of the "3 Speed 2 Inflection Points" Optimum Condition Growth Model

The growth data in this study were 39 ancient pear trees in Lihua Village obtained by specimens of trees, including the annual diameter growth of different types of pear trees. Commonly used tree growth modeling includes the logistic model, the Mitscherlich model [57], the Gompertz model, the Korf model [58], and the Richards model [59]. In the above model fitting process, the sample data will be regarded as a mean tree of the same site and environment. Therefore, it is necessary to propose a growth model, which can not only fit the sample data at different sites and under different environments but also fit the overall growth trend of the sample data. Tree growth consists of three basic processes, namely cell division, cell elongation, and cell differentiation. Theoretically, the growth potential of cells and tissues is unlimited, and their growth should always be exponential. However, since the internal interactions between individual cells or organs limit growth [60], the growth process of trees is divided into the juvenile period, the medium period, and the near-mature period. In this study, the three stages were summarized as "3 speed 2 inflection points," and the overall growth trend was relatively stable. Different tree growth indexes were assigned to each sample data, which is the composite index of site index, structure index, and growth rate index. By selecting the optimal tree growth index and the tree growth model, the optimal tree growth can be obtained, as shown in Formula (4).

$$\begin{cases} d_{1.3} = a_1 \cdot e^{\frac{-b_1}{t}} \cdot \left(e^{-b_4 \cdot t + a_4} + a_5 \right) & 0 < t \leq t_1 \\ d_{1.3} = a_1 \cdot e^{\frac{-b_2}{t} + a_2} \cdot \left(e^{-b_4 \cdot t + a_4} + a_5 \right) & t_1 < t \leq t_2 \\ d_{1.3} = a_1 \cdot e^{\frac{-b_3}{t} + a_3} \cdot \left(e^{-b_4 \cdot t + a_4} + a_5 \right) & t_2 < t \end{cases} \tag{4}$$

$d_{1.3}$ is the DBH, t is the age of the tree, a_1, a_2, a_3, b_1, b_2, and b_3 are the parameters of the tree growth model of the overall sample data. $e^{-b_4 \cdot t + a_4} + a_5$ is the tree growth index model of each independent sample data, and b_4, a_4, and a_5 are the factors of this model.

3. Results

3.1. Experiment Preparation

We choose Beicao Village (Figure A1), Gaodian Village (Figure A3), Houyechang Village (Figure A3), Nandi Village (Figure A1), Qiancao Village (Figure A1), Shilipu Village (Figure A2), Taiziwu Village (Figure A2), Xinzhuang Village (Figure A2), Zhouying Village (Figure A4), and Zhuzhuang Village (Figure A4) as the study area. The parameters of the pear tree were obtained by using an electronic total station, diameter tape, and an increment-borer, and the system accuracy was verified.

3.2. Analysis of the Improved UAV Photogrammetric Program

According to the computer three-dimensional visual algorithm, the matched points are handled to generate the sparse point cloud (Figure 10). The sparse point cloud is encrypted to obtain the dense point cloud with a geographical reference, as shown in Figure 11. The missing point cloud problem is the most common phenomenon in the photogrammetric system. The main solution of this study is to use missing area images separately for three-dimensional point cloud construction and select the high-precision and high-density point cloud construction options of Pix4Dmapper. The precision results of orientation points and check points by using the bundle adjustment are shown in Table 4.

Analysis of results from different data sources shows that the RMSE in the orientation point and check point of Data Source A was generally less than that of Data Source B. The RMSE in the plane of Data Source B is more than two times higher than that of Data Source A, and the RMSE in the elevation Data Source B is also more than two times higher than that of Data Source A.

Figure 10. Sparse 3D point cloud of ancient tree communities.

Figure 11. Dense 3D point cloud of ancient tree communities.

Table 4. Aerial triangulation accuracy for different data sources.

Data Sources		Data Source A	Data Source B
Fundamental orientation points	RMSE of plane	0.233 m	0.442 m
	Maximum error of plane	0.317 m	0.762 m
	RMSE of elevation	0.154 m	1.516 m
	Maximum error of elevation	0.196 m	2.289 m
Check points	RMSE of plane	0.299 m	0.314 m
	Maximum error of plane	0.691 m	0.714 m
	RMSE of elevation	0.873 m	1.508 m
	Maximum error of elevation	1.498 m	2.686 m

3.3. Analysis of the Tree Information Extraction Method

After multiple processing experiments, tree information extraction has the best processing effect by dividing areas (no more than 10 km^2). The point cloud data is displayed by the elevation of ancient tree communities, as shown in Figure 12. The classification results of ground points are shown in Figure 13. The DSM (Figure 14a), DEM (Figure 14b), and CHM (Figure 14c) after processing are shown in Figure 14.

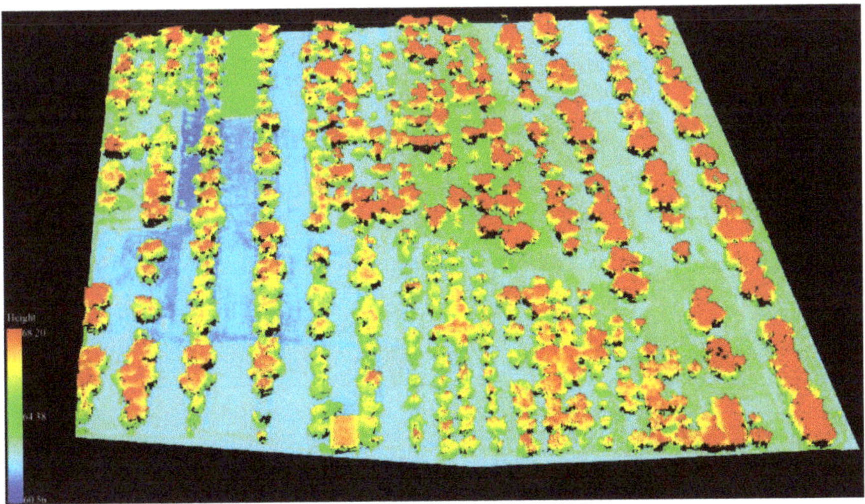

Figure 12. Point cloud data displayed by elevation of ancient tree communities.

Figure 13. Ground point classification of ancient tree communities.

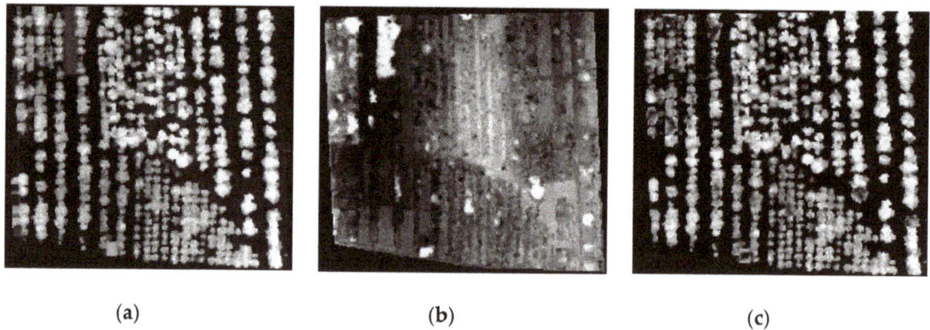

(a)   (b)   (c)

Figure 14. The DSM (a), DEM (b), and canopy height model (CHM) (c) of ancient tree communities.

According to the seed points (Figure 15), the prominent position of point cloud data is segmented and the information of individual trees is calculated, as shown in Figure 16.

To verify the measurement accuracy of tree height in the system, the reference tree heights of 745 pear trees were arranged from small to large, and the tree number was reset. The tree height was distributed between 3.02 and 5.42 m (Figure 17). The results (Table 5) show that the measured values were distributed on both sides of the reference values, and the maximum measurement error was 0.688 m. Most of the measurement error was within 0.3 m.

Table 5. Verification and analysis of measurement accuracy of tree height and crown width.

	RMSE	RMSE%	Bias	Bias%
Tree Height (m)	0.1814 m	4.39%	0.0408 m	0.99%
Crown width (m)	0.3292 m	4.73%	0.0244 m	0.35%

In order to verify the measurement accuracy of the crown width of the system, the reference values of 745 pear trees were arranged according to the size, and the tree number was reset. The canopy amplitude was distributed between 3.01 and 12.02 m (Figure 18). The results (Table 5) show that the measured values are distributed on both sides of the reference value, with the maximum measurement error of 1.165 m. Most of the measurement errors were within 0.5 m.

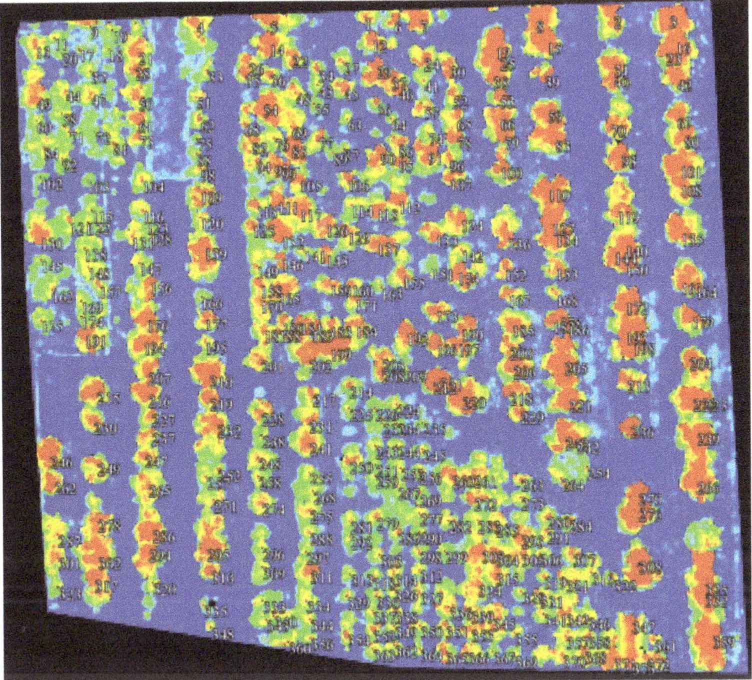

Figure 15. Seed point generated in ancient tree communities.

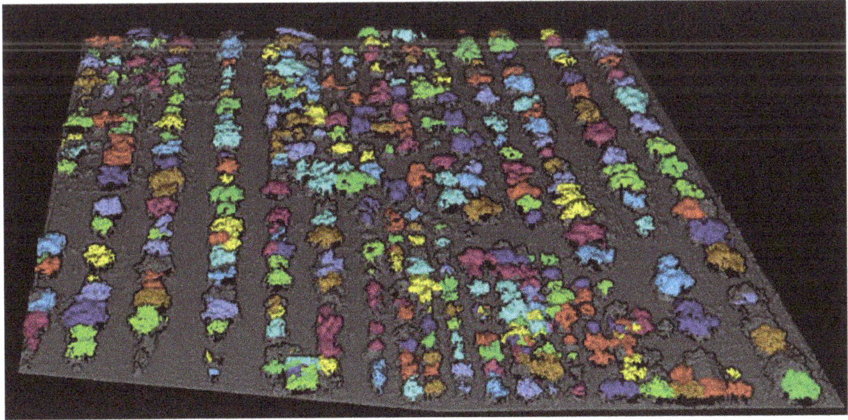

Figure 16. Eye-dome lighting (EDL) display of ancient tree communities.

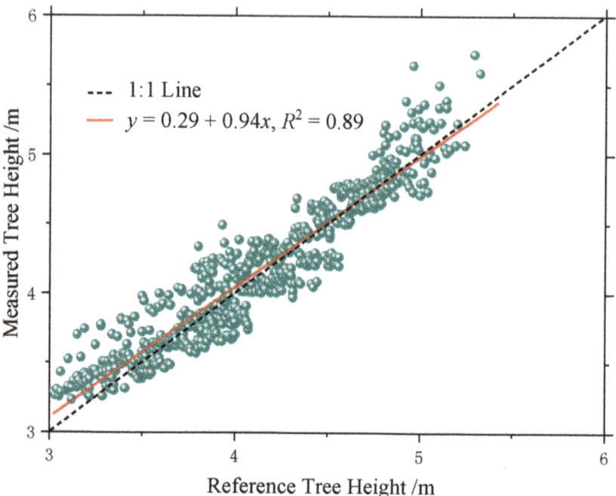

Figure 17. Reference value and measured value distribution of tree height.

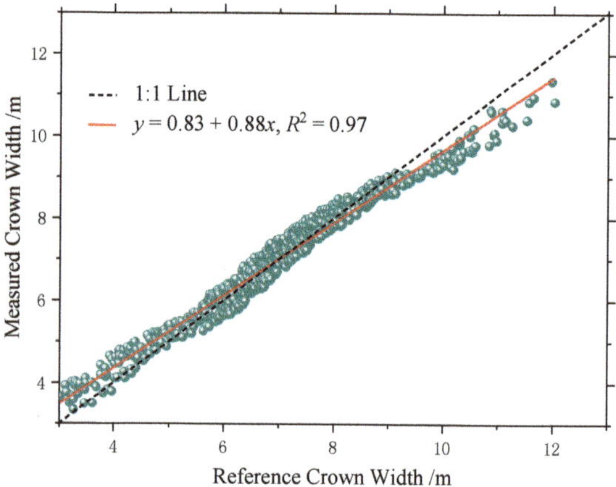

Figure 18. Reference value and measured value distribution of tree crown width.

3.4. Analysis of the Ancient Tree DBH and Age Prediction Model

The fitting status of the height–crown-width–DBH model is shown in Table 6. In order to verify the DBH prediction accuracy of the system, the reference values of 745 pear tree data was adjusted according to the size, and the tree number was reset. DBH was distributed between 16.7 and 55.3 cm (Figure 19). The results (Table 7) show that the predicted values were distributed on both sides of the reference value, and the maximum prediction error was 10.33 cm. Most of the prediction errors were within 5 cm.

Table 6. Fitting analysis of the height–crown-width– diameter-at-breast-height (DBH) model ($R^2 = 0.986$).

Factor	Estimated Value	Standard Error	Lower Limit of 95% Confidence Interval	Upper Limit of 95% Confidence Interval
g_1	1.570	0.193	1.191	1.950
q_1	1.428	0.052	1.325	1.530
g_2	2.296	0.242	1.821	2.771
q_2	1.119	0.035	1.051	1.187

Table 7. Accuracy analysis of DBH prediction.

	RMSE	RMSE%	Bias	Bias%
DBH (cm)	3.0039 cm	9.25%	−0.4193 cm	−1.29%

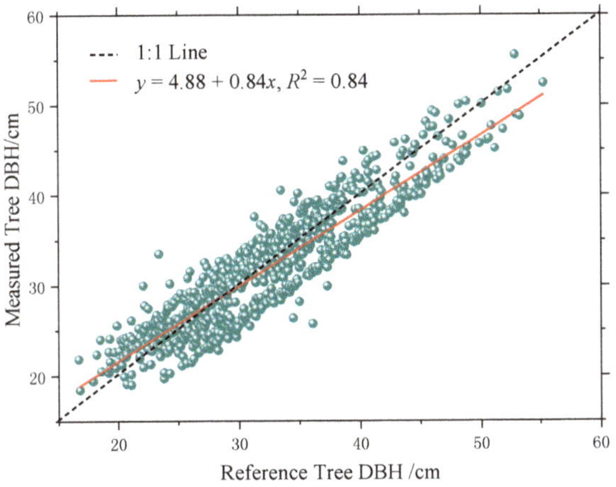

Figure 19. Reference value and measured value distribution of DBH.

In the "3 speed 2 inflection points" model, K-means cluster analysis was carried out on the sample data of 39 ancient pear trees, which was divided into three growth stages: 1–26 years as the juvenile period, 27–59 years as the medium period, and ≥60 years as the near-mature period. In addition, the sample of the optimal tree growth index was taken as the mean tree, and the fitting analysis of the "3 speed 2 inflection points" model is shown in Table 8.

Table 8. Fitting analysis of "3 speed 2 inflection points" model ($R^2 = 0.892$).

Factor	Estimated Value	Standard Value	Lower Limit of 95% Confidence Interval	Upper Limit of 95% Confidence Interval
a_1	26.959	0.747	25.494	28.424
b_1	12.076	0.455	11.184	12.969
a_2	0.522	0.035	0.453	0.590
b_2	25.907	0.866	24.208	27.605
a_3	1.341	0.054	1.236	1.447
b_3	74.661	3.330	68.131	81.191
a_4	5.739	1.778	2.207	9.272
b_4	1.326	0.521	0.290	2.361
a_5	1.147	0.006	1.135	1.159

In order to verify the accuracy of the system's age prediction, the reference ages of 745 pear trees were arranged from small to large, and the tree number was reset, and the ages were distributed between 30 and 104 (Figure 20). The results (Table 9) show that the predicted value is distributed on both sides of the reference value, and the maximum prediction error is 16 years. Most of the prediction errors were within 8 years. The number of pear trees identified through the system was 391, and the number of pear trees sampled through the increment borer was 401.

Table 9. Age prediction accuracy analysis.

	RMSE	RMSE%	Bias	Bias%
Tree Age/years	4.3753 years	8.77%	1.5141 years	3.03%

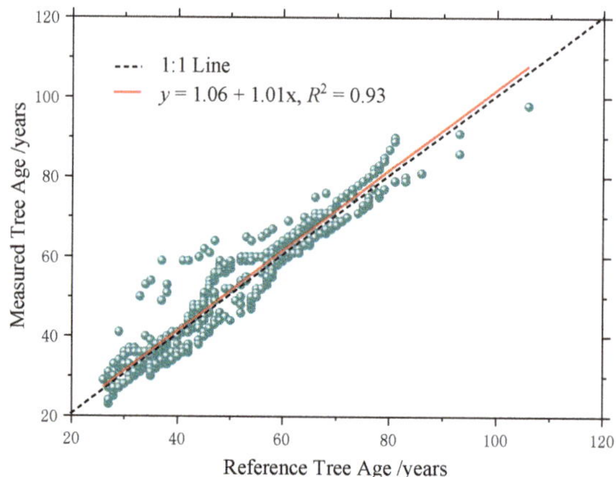

Figure 20. Reference value and measured value distribution of tree age.

4. Discussion

4.1. Comparative Analysis of UAV Photogrammetry in Forestry Application

4.1.1. UAV Photogrammetry to Obtain Forest Structure

There are many studies on extracting the scale structure information of single trees by UAV photogrammetry. Dandois and others (2010) used cameras mounted on a kite platform to obtain aerial images of different-age forests and same-age forest and reconstructed 3D point clouds through Ecosynth. Image construction CHM can be used to estimate tree height ($R^2 > 0.64$) [61]. However, the above research differs greatly from the accuracy of the estimated tree height by our improved system, mainly because our fixed-wing UAV platform is relatively stable. Zarco-Tejada and others analyzed the near-infrared images of olive trees obtained by fixed-wing UAVs, conducted 3D construction by Pix4UAV software (Swiss company Pix4D), and obtained tree height information from DSM image construction, which had good correlation with ground measured trees ($R^2 = 0.83$, RMSE = 35 cm) [62]. The above research is very close to the accuracy of the tree height estimation of this system, but there is still a certain gap in accuracy. The main reason is that our system matches and corrects UAV image and POS data before 3D construction, which is also an innovative and important means to improve the accuracy of low-cost UAV photogrammetry. Ni and others obtained northern forest aerial images through a multi-rotor UAV, reconstructed the three-dimensional point cloud images with Agisoft Photoscan, compared and analyzed the photogrammetric CHM and the LiDAR CHM, and found that the soil scale forest was highly correlated ($R^2 = 0.87$, RMSE = 1.9 m) [63]. In the above research, a multi-rotor UAV platform is used, so there is still a certain gap compared to the fixed-wing UAV platform accuracy. In addition, White and others used RSG to perform the three-dimensional construction of high-altitude aerial images (point cloud density is 12.27 points/m^2) [64]. Because the side overlap rate is too low, only fore-and-aft overlap images are used to match. Image construction point cloud minus LiDAR DEM obtains a terrain-normalized point cloud and performs hierarchical analysis according to the slope and canopy closure change. The image construction point cloud and the LiDAR point cloud feature quantities are statistically significant, and the DBH is the largest difference. The model result is the same. There is no trend between slope and canopy closure. However, the combination of photogrammetry and LiDAR proposed by the institute provides a direction for the future extraction of forest structures with higher precision.

4.1.2. UAV Photogrammetry to Obtain Topography under Tree Crowns

There are many related studies on UAV photography to measure the terrain under the forest. Dandois and others (2010) used aerial images of different forests and same-age forests for 3D construction [61]. By comparing the image construction point cloud DEM with the LiDAR DEM, it was found that the image construction DEM accuracy is low due to the influence of forest canopy occlusion. Dandois and others (2013) used three-dimensional construction to obtain aerial images of deciduous forest growing seasons and deciduous seasons [65]. By comparing the image construction DEM and the LiDAR DEM in the growing season and the deciduous season, it was found that the DEM accuracy (RMSE is 0.89–3.04 m) in the deciduous season images is higher than that in the growing season (RMSE is 2.49–5.69 m). In the difference value between the image construction DEM and the LiDAR DEM, the forest coverage area has a larger DEM difference value than the non-forest area. Wallace and others used the aeronautical aerial image of Eucalyptus forest for 3D construction. The image construction DEM and the LiDAR DEM generally have little difference (the average difference value is 0.09 m). For the high canopy density area, the image construction DEM is not as accurate as the LiDAR DEM. In view of the investigation of ancient tree community, our system chose to take advantage of UAV photogrammetry to extract DEM. The main reason besides the low cost is that the overall density of the ancient pear tree community (7000-60,000 trees/km^2) is small, and the surrounding bare ground is extensive. As a result, the accuracy of image construction DEM is higher.

4.2. Error Analysis of Improved UAV Photogrammetric Measurement System

UAV photogrammetric measurement system is based on studies of SfM-MVS. In the case of tree height and canopy width measurement, errors are caused by all sorts of reasons. As point clouds are scaled, shifted, and rotated into geographic coordinates, the measurement error of each GCP's three-dimensional position will also bring additional registration errors to SfM-MVS. The objective of the modified UAV image matching and correction algorithm in this paper is to minimize the registration errors of SfM-MVS. SfM-MVS can generate data equivalent to TLS over short distances. However, as the measurement distance increases, the accuracy decreases significantly, which is also the main precision limiting factor of UAV photogrammetric tree measurement system. Reasonable flight height can improve the measurement accuracy.

The difference of contrast and terrain texture can also affect the measurement accuracy of the UAV photogrammetric tree measurement system. SfM-MVS also faces many challenges in the vegetation survey, such as the dynamic nature of vegetation, the complexity of vegetation, and the need of many topographic models to filter out vegetation and restore bare surface spots. Although TLS data can also restore vegetation surfaces, most verified data sets are bare land elevation obtained from TS (total station), differential GPS (global positioning system), and other measurement methods. A UAV photogrammetric tree measurement system uses a variety of vegetation filtering algorithms, classifies pixels according to RGB values with multi-scale dimension standards, resampling point cloud at lower resolution, and finally extracts the minimum observation height in a wide and dense vegetation area. In this process of vegetation filtration, the error will increase with the increase of vegetation density. Because of the low stand density of ancient trees, the error is relatively small in the investigation of the ancient tree community. Moreover, for the purpose of protecting the ancient tree community, the Daxing District Gardening and Greening Bureau regularly processed the other vegetation around the ancient tree. Ancient trees, other trees, and bare land are obviously different from each other in images. Therefore, the UAV photogrammetric tree measurement system can be well used for terrain data extraction.

In addition, due to the large imaging distortion residuals of consumer-grade camera, the aerial triangulation calculation of GCPs may lead to the cumulative diffusion of orientation error. Therefore, choosing medium format cameras, high-precision UAV platforms, and optimizing algorithms can not only significantly reduce the uncertainty of the elements of exterior orientation, but also improve measurement accuracy.

4.3. Error Analysis of the Ancient Tree DBH and Age Prediction Model

Both the growth of DBH and changes in tree rings were influenced by many factors such as climate, environment, and the physiological characteristics of vegetation, especially sensitive to climate change. In the past, the non-linear growth model of trees did not consider factors such as climate and environment. The main reason was that the data source was normal-growth trees, and the purpose was to study the growth curve of the tree. Since ancient pear trees are old, it is difficult to study a growth model by a normal-growth trees method. Forest modeling in this study was combined with UAV photogrammetry to obtain a universal model of the growth of ancient pear trees for the inversion of DBH and tree age. Therefore, by effectively classifying the growth index of each pear tree in the data source, the height–crown-width–DBH model and the "3 Speed 2 Inflection Points" optimum condition growth model can simulate the optimal growth process of ancient pear trees.

There are some uncertainties in the establishment of an ancient tree community model, which is caused by many aspects. There are few studies on the physiological parameters of pear species in China, which makes it unreliable to predict the growth trend of pear trees with their physiological parameters. The model is randomly fitted with its own parameters, which results in a certain amount of error in the accuracy of the simulation results. Due to the lack of climate data in this study, there is no relevant climate data to study the optimal growth process of ancient pear trees, which will reduce the accuracy of the model. In addition, forest modeling in this study also has some limitations, and factors such as external interference activities of the forest ecosystem (such as natural disasters, diseases, and pests) are not taken into account. The improvement of model parameters, the physiological parameters of vegetation, and climate data will increase the accuracy of the model simulation.

The forest modeling experiment results in this paper show that most of the predicted results have good consistency and achieve significant relevant predictions. The variability of individual prediction results is mainly caused by individual differences and special environmental impacts. In addition, the distribution of pear community is relatively scattered, and canopy breadth is less affected by forest density. Therefore, the height–crown-width–DBH model's precision is high. However, some pear trees are affected by site conditions and disease and insect disasters, which produces large prediction errors. Because a small number of pear trees grow in sandy environments, this results in slow DBH growth. Therefore, the predicted age is much smaller than the actual age, but most pear trees have better accuracy in predicting age. The overall level of precision could meet the demand of the ancient tree community survey.

5. Conclusions

In this study, a UAV photogrammetric measurement system was developed for the investigation of ancient tree communities. Through an improved UAV photogrammetric program and an ancient tree information extraction method, highly efficient and highly precise measurement of tree height and crown width can be achieved. The accurate prediction of DBH and age can be achieved through the construction of a height–crown-width–DBH model and a "3 speed 2 inflection points" optimum condition growth model.

Although the system is aimed at the investigation of ancient tree communities, the improved UAV photogrammetric program and the ancient tree information extraction method can provide a new method for the application of UAV photogrammetry in a forestry survey. In addition, the height–crown-width–DBH model can predict tree diameter with high precision. The study proposed that a "3 speed 2 inflection points" optimum condition growth model based on tree growth index could predict tree age accurately and provide a new method for age identification in a forest survey. Therefore, a UAV photogrammetric tree measurement system can be applied in ancient tree community surveys and even partial forestry surveys. In the future, if forest environment ground monitoring data can be combined, forest modeling research can be improved and achieve the low-cost and high-precision measurement of stand density, stand volume, carbon storage, and biomass.

Author Contributions: Z.Q. and Z.-K.F. conceived and designed the experiments; Z.Q. and M.W. performed the experiments; Z.Q., Z.L., and C.L. analyzed the data; Z.Q. wrote the main manuscript. All authors contributed in writing and discussing the paper.

Funding: This work was supported by the Medium-to-Long-Term Project of Young Teachers' Scientific Research in Beijing Forestry University (Grant number 2015ZCQ-LX-01), the National Natural Science Foundation of China (Grant number U1710123), and the Beijing Municipal Natural Science Foundation (Grant number 6161001).

Conflicts of Interest: The authors declare no conflict of interest.

Appendix A

Ancient trees are mainly distributed in Panggezhuang Town (Figure A1), Yufa Town (Figure A2), Anding Town (Figure A3), and Zhangziying Town (Figure A4).

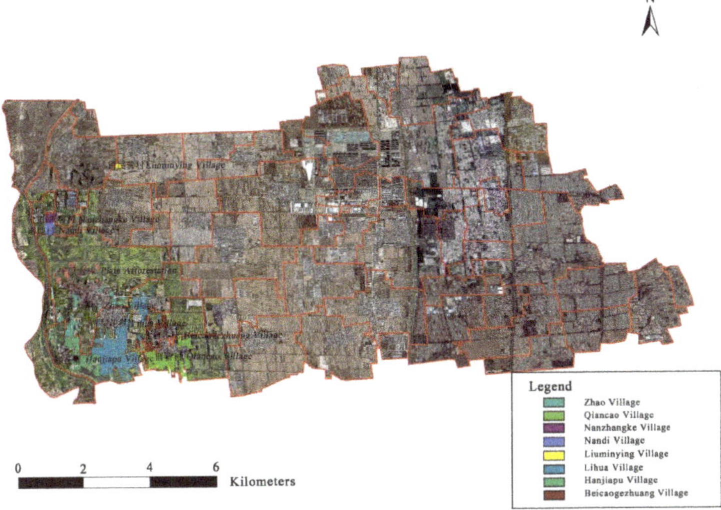

Figure A1. Ancient trees distributed in the Panggezhuang Town.

Figure A2. Ancient trees distributed in the Yufa Town.

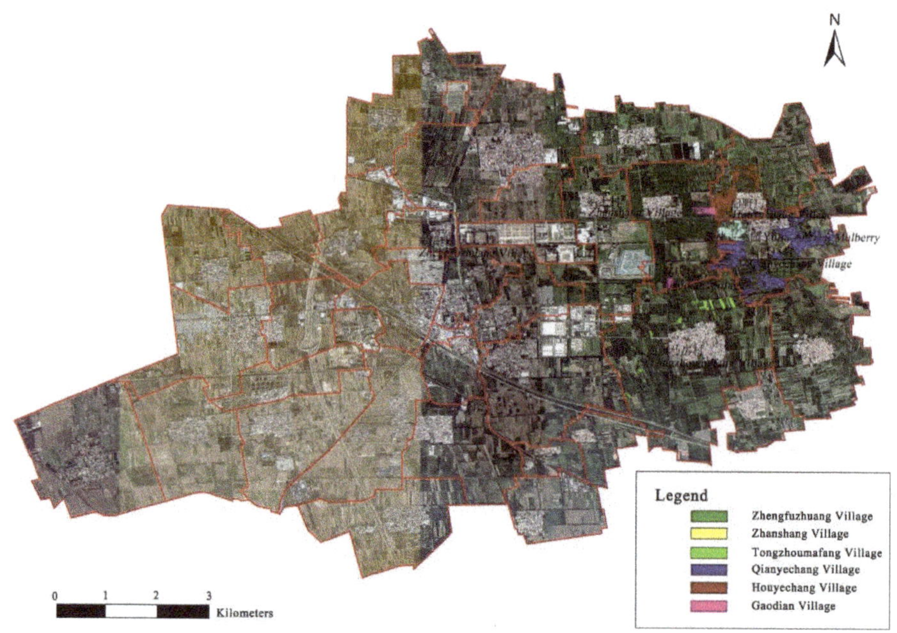

Figure A3. Ancient trees distributed in the Anding Town.

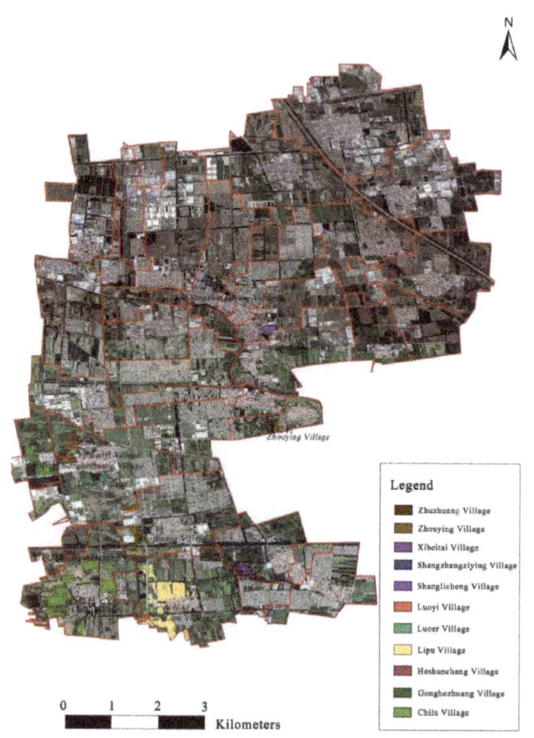

Figure A4. Ancient trees distributed in the Zhangziying Town.

References

1. Brawn, J.D. Effects of restoring oak savannas on bird communities and populations. *Conserv. Biol.* **2006**, *20*, 460–469. [CrossRef] [PubMed]
2. Gibbons, P.; Lindenmayer, D.; Fischer, J.; Manning, A.; Weinberg, A.; Seddon, J.; Ryan, P.; Barrett, G. The future of scattered trees in agricultural landscapes. *Conserv. Biol.* **2008**, *22*, 1309–1319. [CrossRef] [PubMed]
3. Andersson, M.; Milberg, P.; Bergman, K.-O. Low pre-death growth rates of oak (*Quercus robur* L.)—Is oak death a long-term process induced by dry years? *Ann. For. Sci.* **2011**, *68*, 159–168. [CrossRef]
4. Lindenmayer, D.B.; Laurance, W.F.; Franklin, J.F. Global decline in large old trees. *Science* **2012**, *338*, 1305–1306. [CrossRef] [PubMed]
5. Buse, J.; Assmann, T.; Friedman, A.L.L.; Rittner, O.; Pavlicek, T. Wood-inhabiting beetles (coleoptera) associated with oaks in a global biodiversity hotspot: A case study and checklist for israel. *Insect Conserv. Divers.* **2013**, *6*, 687–703. [CrossRef]
6. Helama, S.; Vartiainen, M.; Kolström, T.; Peltola, H.; Meriläinen, J. X-ray microdensitometry applied to subfossil tree-rings: Growth characteristics of ancient pines from the southern boreal forest zone in finland at intra-annual to centennial time-scales. *Veg. Hist. Archaeobot.* **2008**, *17*, 675–686. [CrossRef]
7. Briffa, K.R. Annual climate variability in the holocene: Interpreting the message of ancient trees. *Quat. Sci. Rev.* **2000**, *19*, 87–105. [CrossRef]
8. Foody, G.M.; Atkinson, P.M.; Gething, P.W.; Ravenhill, N.A.; Kelly, C.K. Identification of specific tree species in ancient semi-natural woodland from digital aerial sensor imagery. *Ecol. Appl.* **2005**, *15*, 1233–1244. [CrossRef]
9. Qiu, Z.; Feng, Z.; Jiang, J.; Lin, Y.; Xue, S. Application of a continuous terrestrial photogrammetric measurement system for plot monitoring in the Beijing Songshan National Nature Reserve. *Remote Sens.* **2018**, *10*, 1080. [CrossRef]
10. Yan, F.; Ullah, M.R.; Gong, Y.; Feng, Z.; Chowdury, Y.; Wu, L. Use of a no prism Electronic Total Station for field measurements in *Pinus tabulaeformis* Carr. Stands in china. *Biosyst. Eng.* **2012**, *113*, 259–265. [CrossRef]
11. Qiu, Z.; Feng, Z.; Lu, J.; Sun, R. Design and experiment of forest telescope intelligent dendrometer. *Trans. Chin. Soc. Agric. Mach.* **2017**, *48*, 202–207, 213.
12. Qiu, Z.; Feng, Z.; Jiang, J.; Fan, Y. Design and experiment of forest intelligent surveying and mapping instrument. *Trans. Chin. Soc. Agric. Mach.* **2017**, *48*, 179–187.
13. Van Leeuwen, M.; Nieuwenhuis, M. Retrieval of forest structural parameters using lidar remote sensing. *Eur. J. For. Res.* **2010**, *129*, 749–770. [CrossRef]
14. Liang, X.; Kankare, V.; Hyyppä, J.; Wang, Y.; Kukko, A.; Haggrén, H.; Yu, X.; Kaartinen, H.; Jaakkola, A.; Guan, F. Terrestrial laser scanning in forest inventories. *ISPRS J. Photogramm. Remote Sens.* **2016**, *115*, 63–77. [CrossRef]
15. Murphy, G.E.; Acuna, M.A.; Dumbrell, I. Tree value and log product yield determination in radiata pine (*Pinus radiata*) plantations in Australia: Comparisons of terrestrial laser scanning with a forest inventory system and manual measurements. *Can. J. For. Res.* **2010**, *40*, 2223–2233. [CrossRef]
16. Pueschel, P.; Newnham, G.; Rock, G.; Udelhoven, T.; Werner, W.; Hill, J. The influence of scan mode and circle fitting on tree stem detection, stem diameter and volume extraction from terrestrial laser scans. *ISPRS J. Photogramm. Remote Sens.* **2013**, *77*, 44–56. [CrossRef]
17. Ryding, J.; Williams, E.; Smith, M.J.; Eichhorn, M.P. Assessing handheld mobile laser scanners for forest surveys. *Remote Sens.* **2015**, *7*, 1095–1111. [CrossRef]
18. Nord-Larsen, T.; Schumacher, J. Estimation of forest resources from a country wide laser scanning survey and national forest inventory data. *Remote Sens. Environ.* **2012**, *119*, 148–157. [CrossRef]
19. Holmgren, J.; Persson, Å. Identifying species of individual trees using airborne laser scanner. *Remote Sens. Environ.* **2004**, *90*, 415–423. [CrossRef]
20. Lim, K.; Treitz, P.; Baldwin, K.; Morrison, I.; Green, J. Lidar remote sensing of biophysical properties of tolerant northern hardwood forests. *Can. J. Remote Sens.* **2003**, *29*, 658–678. [CrossRef]
21. Lim, K.S.; Treitz, P.M. Estimation of above ground forest biomass from airborne discrete return laser scanner data using canopy-based quantile estimators. *Scand. J. For. Res.* **2004**, *19*, 558–570. [CrossRef]

22. Næsset, E. Effects of different flying altitudes on biophysical stand properties estimated from canopy height and density measured with a small-footprint airborne scanning laser. *Remote Sens. Environ.* **2004**, *91*, 243–255. [CrossRef]

23. Patenaude, G.; Hill, R.; Milne, R.; Gaveau, D.; Briggs, B.; Dawson, T. Quantifying forest above ground carbon content using lidar remote sensing. *Remote Sens. Environ.* **2004**, *93*, 368–380. [CrossRef]

24. Leberl, F.; Irschara, A.; Pock, T.; Meixner, P.; Gruber, M.; Scholz, S.; Wiechert, A. Point clouds. *Photogramm. Eng. Remote Sens.* **2010**, *76*, 1123–1134. [CrossRef]

25. Bohlin, J.; Wallerman, J.; Fransson, J.E. Forest variable estimation using photogrammetric matching of digital aerial images in combination with a high-resolution dem. *Scand. J. For. Res.* **2012**, *27*, 692–699. [CrossRef]

26. Järnstedt, J.; Pekkarinen, A.; Tuominen, S.; Ginzler, C.; Holopainen, M.; Viitala, R. Forest variable estimation using a high-resolution digital surface model. *ISPRS J. Photogramm. Remote Sens.* **2012**, *74*, 78–84. [CrossRef]

27. Chiang, K.-W.; Tsai, M.-L.; Chu, C.-H. The development of an uav borne direct georeferenced photogrammetric platform for ground control point free applications. *Sensors* **2012**, *12*, 9161–9180. [CrossRef] [PubMed]

28. Fonstad, M.A.; Dietrich, J.T.; Courville, B.C.; Jensen, J.L.; Carbonneau, P.E. Topographic structure from motion: A new development in photogrammetric measurement. *Earth Surf. Process. Landf.* **2013**, *38*, 421–430. [CrossRef]

29. Dalponte, M.; Bruzzone, L.; Gianelle, D. Fusion of hyperspectral and lidar remote sensing data for classification of complex forest areas. *IEEE Trans. Geosci. Remote Sens.* **2008**, *46*, 1416–1427. [CrossRef]

30. An, N.; Welch, S.M.; Markelz, R.J.C.; Baker, R.L.; Palmer, C.M.; Ta, J.; Maloof, J.N.; Weinig, C. Quantifying time-series of leaf morphology using 2d and 3d photogrammetry methods for high-throughput plant phenotyping. *Comput. Electron. Agric.* **2017**, *135*, 222–232. [CrossRef]

31. Kim, D.; Yun, H.S.; Jeong, S.; Kwon, Y.; Kim, S.; Lee, W.S.; Kim, H. Modeling and testing of growth status for Chinese cabbage and white radish with UAV based rgb imagery. *Remote Sens.* **2018**, *10*, 563. [CrossRef]

32. Kane, V.R.; North, M.P.; Lutz, J.A.; Churchill, D.J.; Roberts, S.L.; Smith, D.F.; McGaughey, R.J.; Kane, J.T.; Brooks, M.L. Assessing fire effects on forest spatial structure using a fusion of landsat and airborne lidar data in yosemite national park. *Remote Sens. Environ.* **2014**, *151*, 89–101. [CrossRef]

33. Zellweger, F.; Braunisch, V.; Baltensweiler, A.; Bollmann, K. Remotely sensed forest structural complexity predicts multi species occurrence at the landscape scale. *For. Ecol. Manag.* **2013**, *307*, 303–312. [CrossRef]

34. Hill, A.; Breschan, J.; Mandallaz, D. Accuracy assessment of timber volume maps using forest inventory data and lidar canopy height models. *Forests* **2014**, *5*, 2253–2275. [CrossRef]

35. St-Onge, B.; Audet, F.-A.; Bégin, J. Characterizing the height structure and composition of a boreal forest using an individual tree crown approach applied to photogrammetric point clouds. *Forests* **2015**, *6*, 3899–3922. [CrossRef]

36. Sankey, T.T.; McVay, J.; Swetnam, T.L.; McClaran, M.P.; Heilman, P.; Nichols, M. Uav hyperspectral and lidar data and their fusion for arid and semi-arid land vegetation monitoring. *Remote Sens. Ecol. Conserv.* **2018**, *4*, 20–33. [CrossRef]

37. Sankey, T.; Donager, J.; McVay, J.; Sankey, J.B. Uav lidar and hyperspectral fusion for forest monitoring in the southwestern USA. *Remote Sens. Environ.* **2017**, *195*, 30–43. [CrossRef]

38. Corona, P.; Fattorini, L. Area-based lidar-assisted estimation of forest standing volume. *Can. J. For. Res.* **2008**, *38*, 2911–2916. [CrossRef]

39. Steinmann, K.; Mandallaz, D.; Ginzler, C.; Lanz, A. Small area estimations of proportion of forest and timber volume combining lidar data and stereo aerial images with terrestrial data. *Scand. J. For. Res.* **2013**, *28*, 373–385. [CrossRef]

40. Næsset, E. Predicting forest stand characteristics with airborne scanning laser using a practical two-stage procedure and field data. *Remote Sens. Environ.* **2002**, *80*, 88–99. [CrossRef]

41. Lisein, J.; Pierrot-Deseilligny, M.; Bonnet, S.; Lejeune, P. A photogrammetric workflow for the creation of a forest canopy height model from small unmanned aerial system imagery. *Forests* **2013**, *4*, 922–944. [CrossRef]

42. Lee, H.; Slatton, K.C.; Roth, B.E.; Cropper, W., Jr. Adaptive clustering of airborne lidar data to segment individual tree crowns in managed pine forests. *Int. J. Remote Sens.* **2010**, *31*, 117–139. [CrossRef]

43. Kaartinen, H.; Hyyppä, J.; Yu, X.; Vastaranta, M.; Hyyppä, H.; Kukko, A.; Holopainen, M.; Heipke, C.; Hirschmugl, M.; Morsdorf, F. An international comparison of individual tree detection and extraction using airborne laser scanning. *Remote Sens.* **2012**, *4*, 950–974. [CrossRef]

44. Li, W.; Guo, Q.; Jakubowski, M.K.; Kelly, M. A new method for segmenting individual trees from the lidar point cloud. *Photogramm. Eng. Remote Sens.* **2012**, *78*, 75–84. [CrossRef]

45. Gonzalez-Benecke, C.; Gezan, S.A.; Samuelson, L.J.; Cropper, W.P.; Leduc, D.J.; Martin, T.A. Estimating *Pinus palustris* tree diameter and stem volume from tree height, crown area and stand-level parameters. *J. For. Res.* **2014**, *25*, 43–52. [CrossRef]

46. Carrer, M.; Urbinati, C. Age-dependent tree-ring growth responses to climate in *Larix decidua* and *Pinus cembra*. *Ecology* **2004**, *85*, 730–740. [CrossRef]

47. Climent, J.; Chambel, M.; Pérez, E.; Gil, L.; Pardos, J. Relationship between heartwood radius and early radial growth, tree age, and climate in *Pinus canariensis*. *Can. J. For. Res.* **2002**, *32*, 103–111. [CrossRef]

48. Othman, M.F.; Shazali, K. Wireless sensor network applications: A study in environment monitoring system. *Procedia Eng.* **2012**, *41*, 1204–1210. [CrossRef]

49. Carrivick, J.L.; Smith, M.W.; Quincey, D.J. *Structure from Motion in the Geosciences*; John Wiley & Sons: Chichester, UK, 2016.

50. Smith, M.; Carrivick, J.; Quincey, D. Structure from motion photogrammetry in physical geography. *Prog. Phys. Geogr.* **2016**, *40*, 247–275. [CrossRef]

51. Mosbrucker, A.R.; Major, J.J.; Spicer, K.R.; Pitlick, J. Camera system considerations for geomorphic applications of sfm photogrammetry. *Earth Surf. Process. Landf.* **2017**, *42*, 969–986. [CrossRef]

52. McLauchlan, P.F.; Jaenicke, A. Image mosaicing using sequential bundle adjustment. *Image Vis. Comput.* **2002**, *20*, 751–759. [CrossRef]

53. Chow, J.C.; Lichti, D.D. Photogrammetric bundle adjustment with self-calibration of the primesense 3d camera technology: Microsoft kinect. *IEEE Access* **2013**, *1*, 465–474. [CrossRef]

54. Mouragnon, E.; Lhuillier, M.; Dhome, M.; Dekeyser, F.; Sayd, P. Generic and real-time structure from motion using local bundle adjustment. *Image Vis. Comput.* **2009**, *27*, 1178–1193. [CrossRef]

55. Di, K.; Liu, Y.; Liu, B.; Peng, M.; Hu, W. A self-calibration bundle adjustment method for photogrammetric processing of chang 'E-2 stereo lunar imagery. *IEEE Trans. Geosci. Remote Sens.* **2014**, *52*, 5432–5442.

56. Schnabel, R.; Wahl, R.; Klein, R. *Efficient Ransac for Point-Cloud Shape Detection*; Computer Graphics Forum; Wiley Online Library: Oxford, UK, 2007; pp. 214–226.

57. Dobbertin, M. Tree growth as indicator of tree vitality and of tree reaction to environmental stress: A review. *Eur. J. For. Res.* **2005**, *124*, 319–333. [CrossRef]

58. Zeide, B. Analysis of growth formulas. *For. Sci.* **1993**, *39*, 594–616.

59. Weiner, J.; Thomas, S.C. The nature of tree growth and the "age-related decline in forest productivity". *Oikos* **2001**, *94*, 374–376. [CrossRef]

60. Moulia, B.; Coutand, C.; Lenne, C. Posture control and skeletal mechanical acclimation in terrestrial plants: Implications for mechanical modeling of plant architecture. *Am. J. Bot.* **2006**, *93*, 1477–1489. [CrossRef] [PubMed]

61. Dandois, J.P.; Ellis, E.C. Remote sensing of vegetation structure using computer vision. *Remote Sens.* **2010**, *2*, 1157–1176. [CrossRef]

62. Zarco-Tejada, P.J.; Diaz-Varela, R.; Angileri, V.; Loudjani, P. Tree height quantification using very high resolution imagery acquired from an unmanned aerial vehicle (uav) and automatic 3d photo-reconstruction methods. *Eur. J. Agron.* **2014**, *55*, 89–99. [CrossRef]

63. Ni, W.; Liu, J.; Zhang, Z.; Sun, G.; Yang, A. Evaluation of UAV-Based Forest Inventory System Compared with Lidar Data. In Proceedings of the 2015 IEEE International Geoscience and Remote Sensing Symposium (IGARSS), Milan, Italy, 26–31 July 2015; pp. 3874–3877.

64. White, J.C.; Stepper, C.; Tompalski, P.; Coops, N.C.; Wulder, M.A. Comparing als and image-based point cloud metrics and modelled forest inventory attributes in a complex coastal forest environment. *Forests* **2015**, *6*, 3704–3732. [CrossRef]

65. Dandois, J.P.; Ellis, E.C. High spatial resolution three-dimensional mapping of vegetation spectral dynamics using computer vision. *Remote Sens. Environ.* **2013**, *136*, 259–276. [CrossRef]

 *forests*

Article

Detection of Coniferous Seedlings in UAV Imagery

Corey Feduck [1], Gregory J. McDermid [1,*] and Guillermo Castilla [2]

[1] Department of Geography, 2500 University Dr. NW Calgary, University of Calgary,
Calgary, AB T2N 1N4, Canada; corey.feduck@ucalgary.ca
[2] Northern Forestry Centre, Canadian Forest Service, 5320 122 St. NW Edmonton,
Victoria, AB T6H 3S5, Canada; guilermo.castilla@canada.ca
* Correspondence: mcdermid@ucalgary.ca; Tel.: +1-403-220-4780

Received: 26 June 2018; Accepted: 16 July 2018; Published: 18 July 2018

Abstract: Rapid assessment of forest regeneration using unmanned aerial vehicles (UAVs) is likely to decrease the cost of establishment surveys in a variety of resource industries. This research tests the feasibility of using UAVs to rapidly identify coniferous seedlings in replanted forest-harvest areas in Alberta, Canada. In developing our protocols, we gave special consideration to creating a workflow that could perform in an operational context, avoiding comprehensive wall-to-wall surveys and complex photogrammetric processing in favor of an efficient sampling-based approach, consumer-grade cameras, and straightforward image handling. Using simple spectral decision rules from a red, green, and blue (RGB) camera, we documented a seedling detection rate of 75.8 % ($n = 149$), on the basis of independent test data. While moderate imbalances between the omission and commission errors suggest that our workflow has a tendency to underestimate the seedling density in a harvest block, the plot-level associations with ground surveys were very high (Pearson's $r = 0.98$; $n = 14$). Our results were promising enough to suggest that UAVs can be used to detect coniferous seedlings in an operational capacity with standard RGB cameras alone, although our workflow relies on seasonal leaf-off windows where seedlings are visible and spectrally distinct from their surroundings. In addition, the differential errors between the pine seedlings and spruce seedlings suggest that operational workflows could benefit from multiple decision rules designed to handle diversity in species and other sources of spectral variability.

Keywords: unmanned aerial vehicles; seedling detection; forest regeneration; reforestation; establishment survey; machine learning; multispectral classification

---

1. Introduction

The rapid assessment of forest and vegetation structure using unmanned aerial vehicles (UAVs) is likely to decrease the cost of field surveys for a variety of resource industries. UAVs may be particularly well suited for applications in reforestation, because they can collect very high-resolution imagery of small features with great operational flexibility. In Canada, establishment surveys are conducted at every forest-harvest area that has been replanted, to assess the adequacy of spacing, survival, growth, and species composition. For example, the Regeneration Standards of Alberta [1] call for a reconnaissance survey to be conducted three growing seasons after planting, wherein certified forestry technicians walk through the harvest area to visually estimate 'stocking', the percentage of 10 m^2 cells within the block that contain a live seedling at least 30 cm in height, from an acceptable tree species. If the estimated stocking rate is above 84%, the harvest area passes the establishment assessment. If the stocking rate is below 70%, the harvest area is rejected and must be replanted. If the stocking falls between 70%–84%, the harvest area becomes subject to further assessment [1].

If the condition, minimum height, and species of seedlings within the sample cells used to perform establishment surveys could be derived from UAV imagery, then the reduced need for manual surveys could lead to considerable cost savings. However, it needs to be demonstrated that the seedlings can be automatically or semi-automatically detected from a remote sensing. Coniferous seedlings under five years of age in Canada have crown diameters between 5 and 30 cm; imagery of a very high spatial resolution is required to detect the seedlings of this size. Although automated procedures have been used to detect small tree stumps [2], and to detect weed seedlings on bare soil in an agricultural application [3], we could not find any previous research attempting to identify coniferous seedlings of this age in an automated manner. Early studies of remote sensing applications in forestry [4,5] used manual photointerpretation of large-scale aerial photographs acquired from piloted helicopters. For example, Hall and Aldred [5] detected just 44% of seedlings with crown diameter less than 30 cm using 1:500 color-infrared photography. More recently, Goodbody et al. [6] classified 2.4 cm spatial resolution red, green, and blue (RGB) imagery acquired from a UAV over harvest blocks replanted 5 to 15 years earlier in British Columbia, Canada, and obtained user accuracies for coniferous cover between 35% and 97%. However, no attempt was made to detect individual conifer seedlings or saplings, which in their study area were greater than 1 m in height. These authors acknowledged that the potential to detect all stems using aerial remote-sensing technologies is still limited, and called for further research.

Depending on the spatial resolution of the imagery, seedling detection is akin to individual-tree detection in mature forests, which has been well studied using satellites [7], piloted aircraft [4,8–15] and, more recently, UAVs [16–23]. Given a fixed spatial resolution, the accuracy with which individual trees are detected tends to improve with crown size [11]. For example, using 15 cm resolution multispectral airborne imagery and an image-segmentation algorithm, Hirschmugl et al. [24] obtained 70% accuracy on replanted coniferous trees between 5 and 10 years of age, with an average height of 138 cm. The accuracy levels improved for the crown diameters larger than 30 cm. The UAV-based studies of the tree detection have achieved even higher accuracies. For example, Wallace et al. [16] obtained overall accuracies of 97% (n = 308) using a tree-crown segmentation algorithm applied to a dense (>50 points/m^2) light detection and ranging (LiDAR) point cloud. However, the LiDAR sensors come with equipment costs and operational difficulties that some may wish to avoid. Consumer-grade cameras mounted on UAVs provide an attractive low-cost source of vegetation information over disturbed and regenerating forests [25,26].

In this research, we show how millimetric (i.e., spatial resolution on the order of millimeters) UAV imagery can be used to detect coniferous seedlings less than five years old within forest-harvest areas in Alberta, Canada, with good accuracies using simple processing workflows. This study represents the first step towards creating a larger UAV-based stocking-assessment workflow, which, once realized, could extend to the remote assessment of height (from LiDAR or photogrammetric point clouds), species, and condition (from deep-learning algorithms). Achieving this complete workflow would reduce the need for in situ assessments of forest stocking, and provide a powerful new tool for establishment surveys.

2. Materials and Methods

2.1. Study Area

We surveyed two replanted forest-harvest areas located in western Alberta, Canada, for this study (Figure 1). One of the harvest areas was used to develop and train the seedling-detection algorithm, and is hereafter referred to as the 'training study area'. A second block was used as an independent validation site and is hereafter referred to as the 'test study area'. The 20.3 ha training study area is managed by Weyerhaeuser Canada (Pembina Forest Management Area), and was replanted approximately four years before our survey, with a mix of lodgepole pine (*Pinus contorta*) and white spruce (*Picea glauca*) seedlings. The 3.3 ha test study area was also replanted with a mix of lodgepole pine and white spruce seedlings between three and four years before our field survey, although most individuals we encountered in the field were lodgepole pine. The vegetation surrounding these two harvest areas consist of forests regenerating to mixed stands of aspen (*Populus tremuloides*), lodgepole pine, and white spruce.

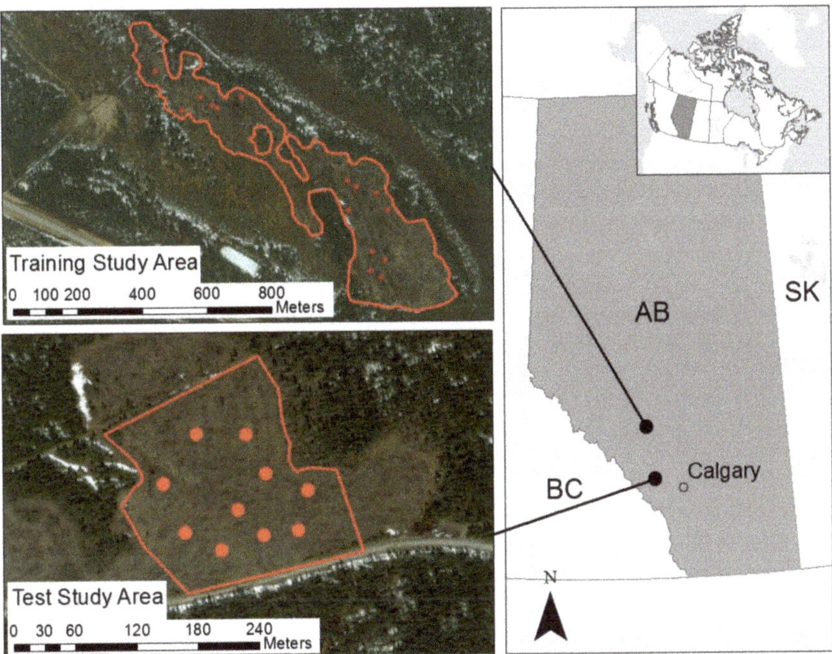

Figure 1. Location of two replanted forest-harvest areas surveyed for this research in southwestern Alberta, Canada. Sample plots within the harvest areas are depicted with the red dots.

2.2. Reference Data

Field crews surveyed the test study area on 24 April 2014, and the training study area on 1 October 2015. We timed our visits to exploit the leaf-off seasonal windows in the spring and fall when the coniferous seedlings have increased their spectral contrast with their surroundings. We contend that the 1.5 year time gap between the two surveys is irrelevant, given that testing took place independently of the training. The crews located randomly generated plot centers using handheld global positioning system (GPS) units and established 50 m^2 circular plots with 3.99 m radii (Figure 2). Biodegradable clay targets or plastic boards were placed in the center of the plot and pinned down with metal spikes as ground control points, whose precise locations were recorded with a survey-grade Trimble real-time kinematic (RTK) global navigation satellite system (GNSS) unit. The plot outlines were marked using chalk or spray paint. The crews then recorded the species and precise location of each seedling inside the plot using the RTK GNSS. A total of 254 seedlings within 14 plots in the training study area, and 149 seedlings within 10 plots in the test study area were surveyed in this manner.

2.3. UAV Imagery

UAV imagery was collected by a flight crew consisting of a pilot and a spotter using a 3DR X8 + octocopter (Figure 3a). Both areas were flown in conjunction with the seedling surveys 24 April 2014 for the test study area and 1 October 2015 for the training study area. Details of the X8+ platform and payload are summarized in Table 1. The platform was modified to carry two cameras simultaneously, one standard RGB camera and a second camera with a modified red-edge (RE) filter. Single-scene images, one RGB and one RE, were acquired for each of the 24 plots during 24 separate flights. The flights took place between 10:00 am and 5:30 pm to cover a variety of lightning conditions. The platform operated a simple automated flight plan, as follows: The X8+ launched, flew to the plot center, and hovered at 15 m above ground level to acquire imagery (single photographs) at a consistent scale (Figure 3b). We installed a LidarLITE laser range finder (vertical accuracy < 2.5 cm) to the UAV, which allowed us to control the altitude of the X8 for imaging. It is important to note that that circular plot did not cover the entire image, but was instead located at the center of each frame. As the UAV hovered directly over each plot prior to image acquisition, the plots were always located very close to the principal point, with the maximum off-nadir angles never exceeding 15 degrees. This reduced the terrain distortion and layover effects. Each flight was less than three minutes in length. The imagery was collected using a sampling approach, avoiding the need for wall-to-wall aerial surveys designed to image the entire harvest area.

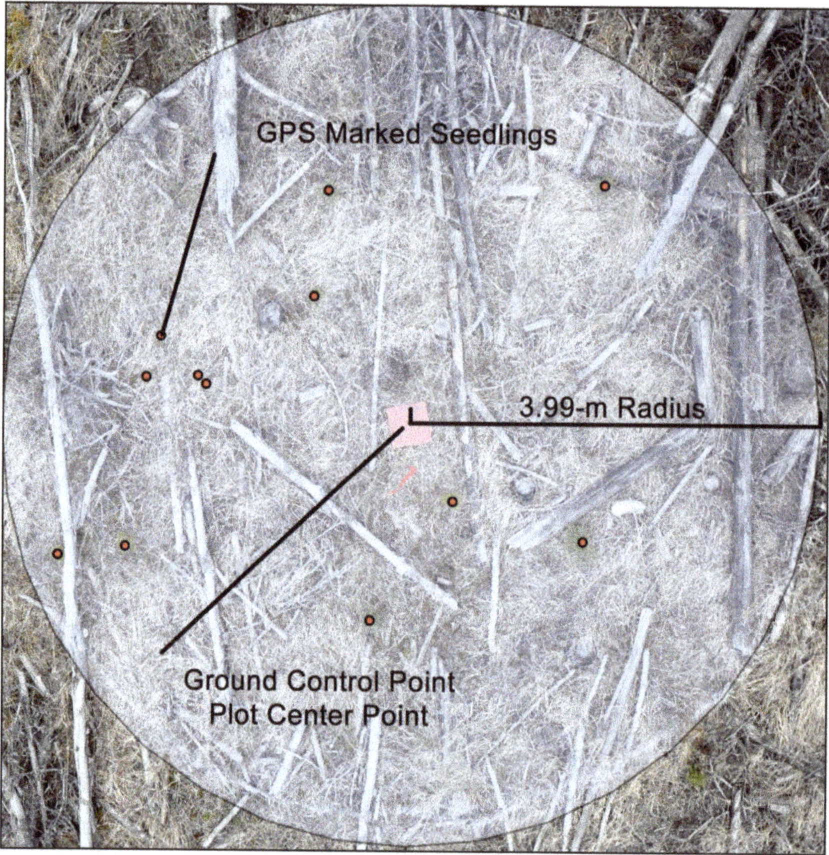

Figure 2. A sample plot, outlined with paint, measured 3.99 m in radius. The coordinates of both ground control point/center point and seedlings were measured using a survey grade global navigation satellite system (GNSS) unit. Note that the image has been clipped to just the plot extent for the purpose of display.

We used a Nikon Coolpix A digital camera to collect standard RGB imagery and a modified Canon PowerShot S110 to collect imagery in the RE wavelength. We substituted the internal near-infrared (NIR) filter in the Canon PowerShot S110 with an Event 38 near-infrared green blue

(NGB) filter, which pushed the red band response to be centered on 715 nm. The aerial imagery was collected at a low altitude of 15 m, resulting in a ground sampling distance of 3 mm for the Nikon Coolpix A (18.5 mm focal length) and 5 mm for the Canon PowerShot S110 (5.2 mm focal length). The cameras were set to a fixed shutter speed of 1/1250 s with varying apertures, and were manually triggered using a remote control.

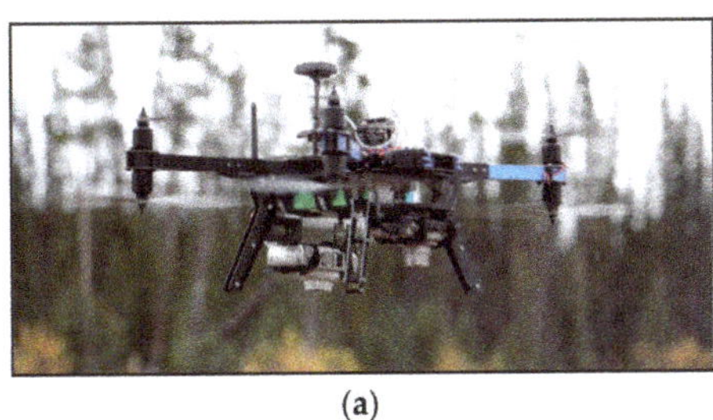
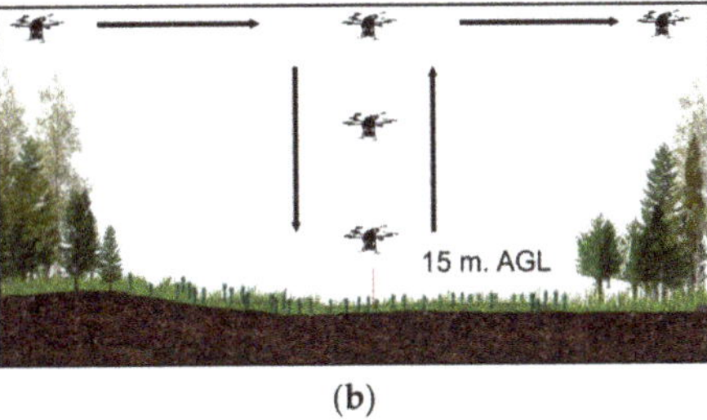

(a) **(b)**

Figure 3. (**a**) The 3DR X8+ unmanned aerial vehicle (UAV) can collect density samples at a rate of 45 s/ha. (**b**) The sample-based flight plan used for this study enables a rapid assessment of large forestry blocks. The UAV moves to a sample plot waypoint, descends to a 15 m altitude above ground level (AGL), captures an image, ascends, and continues to the next waypoint.

Table 1. Unmanned aerial vehicle (UAV) and camera specifications. RGB—red green blue. NIR—near-infrared; CMOS—complementary metal-oxide-semiconductor; NDVI—normalized difference vegetation index.

UAV Specifications: 3DR X8+		
Description	Octocopter	
Vehicle Dimensions	35 cm × 50 cm × 21 cm	
Battery	4S 14 8V 10.000 mAh 10C	
Vehicle Weight with Battery	2.56 kg	
Platform Estimated Flight Time	15 min	
Maximum Speed	96 km/h	
Ranging Device	LidarLITE laser range finder	
Payload Specification	**RGB**	**NIR**
Camera Model	Nikon Coolpix A	Canon PowerShot S110
Weight	299 g	198 g
Image Size (megapixels)	16 MP	12.1 MP
Ground Sampling Distance at Nadir	3 mm	5 mm
Image Dimensions (pixels)	(4928 × 3264)	(4000 × 3000)
Effective Field of View for Sample Plot	30°	30°
Focal Length	18.5 mm	5.2 mm
Aspect Ratio	3:2	4:3
Filter	Stock	Event 38 NDVI
Bit Depth	24	24
Trigger Mode	Shutter Priority	Shutter Priority
ISO	500	500
Shutter Speed	1/1250	1/1250
Maximum Aperture	*f*/4	*f*/3.5
Focus Mode	Center weighted	Center weighted
Sensor Type	CMOS	CMOS

2.4. Data Handling and Image Analysis

The seedling locations were exported to a geographic information system (GIS) point layer, visually confirmed using the UAV imagery, and, if required, spatially edited to be centered within each seedling crown in the image. The seedlings ranged in height from 5 cm to 35 cm, and the crown radii ranged between 5 cm and 50 cm. The tall seedlings were those manually planted about four years before our study, while the short seedlings regenerated naturally. The wider crowns (up to 50 cm) corresponded generally to clusters of naturally regenerating seedlings, rather than to individuals. We treated these clusters as single entities for the purpose of this study. It is important to note that the planted seedlings (taller, generally isolated from other individuals) are more important than the naturally regenerating seedlings (shorter, sometimes occurring in clusters) when assessing stocking in the planted harvest areas. Not all of the seedlings will survive to maturity, and the planted seedlings have the best chance. Within a cluster, no more than one seedling will typically survive.

The image analysis workflow is a three-step object-based process consisting of (i) image segmentation, (ii) automated classification using a classification and regression tree (CART) machine-learning algorithm, and (iii) the merging of adjacent image objects classified as 'seedlings' into single seedling objects. Priority was placed on creating a workflow that is economic with regards to both the UAV flight time and processing time, so image analyses were conducted on single scenes with minimal preprocessing. The challenge was to identify a classification ruleset that could perform under a variety of target and illumination conditions. We used the CART approach in the training study area to test the importance of the spectral, spatial, and textural variables for this task. On the basis of the results of this testing, we selected a single model for application in the test study area.

Red-edge and RGB imagery were co-registered into 16-bit unsigned raster layers for each sample, using ArcGIS Desktop 10.1 (Figure 4a,b). The images were rectified with first-order polynomial functions, using the seedlings and ground control points as reference marks. It is worth noting that this step—rectifying images to match field data—would not be required in an operational workflow. Additional raster layers were generated from band ratios to gain a spectral contrast between the green seedlings and their non-photosynthetic surroundings (Figure 4c). We used Trimble eCognition Suite 9.1 (www.ecognition.com) to segment the imagery and derive image-object statistics. The input raster for the initial segmentation was a ratio of ratios (red ratio/blue ratio) scaled to the 0–255 interval. The user-defined segmentation parameters were as follows: scale = 50; shape = 0.1; and compactness = 0.3. All of the 48 images (24 RGB and 24 RE) were segmented using the same parameters. We arrived at the final segmentation parameters iteratively through trial and error. Our goal was to develop object primitives that best delineated the seedling edges from their surroundings. The resulting image-objects were then further merged using a homogenous region-growing algorithm, with shape and compactness factors of 0.1 and 0.5, respectively. Once the final image objects were generated for each plot, we assembled a number of attributes for each image object. The final list of spectral, spatial, and textural variables evaluated by the CART approach is summarized in Table 2.

2.5. Machine Learning

Image-object attributes were exported to a table, resulting in a database with 18,905 records (image-objects from all of the 14 plots in the training study area together). Each record was then classified as either seedling or non-seedling using the CART machine-learning algorithm in the Salford Predictive Modeler (SPM v. 70) software (info.salford-systems.com). This algorithm generates a classification decision tree, with rules that can be used with structured query language (SQL) queries or to build a decision tree in eCognition. Three sets of models were evaluated. All three of the sets used the same spatial and textural attributes as the predictors (Table 2), but the spectral attributes varied as follows: (i) RGB-only variables, (ii) RE-only variables, and (iii) RGB-combined-with-RE variables. The adjacent image objects classified as seedling were merged together using a GIS 'dissolve' function. The detection accuracy of each model was assessed using a 10-fold cross validation procedure. The mean overall seedling classification accuracy was obtained by each model across all of the 10 trials and was assigned as a measure of global accuracy. The most accurate model was applied to the test study area, which served as an independent validation of our workflow.

(a)   (b)   (c)

Figure 4. Detail of the UAV image from a sample plot. Left to right: (**a**) red, green, and blue (RGB) (3 mm spatial resolution), (**b**) red-edge (RE) (5 mm spatial resolution), and (**c**) red ratio/blue ratio (from the RGB image).

3. Results and Discussion

3.1. Model Selection

The overall classification accuracy of all of the image-objects in the training study area was 96%, 97%, and 97% for the RGB-only variables, RE-only variables, and RGB and RE variables, respectively. It should be noted that the overall accuracy reported here (raw agreement) is a high-level accuracy statistic based on a disproportionately small number of seedling objects (2420) to non-seedling objects (18,663). We report on more detailed error analytics associated with the test dataset below. As the RE-only and RGB and RE models did not result in significant increases in performance (1%), we chose to use the RGB-only model for parsimony.

The final CART decision tree was pruned to a simple two-rule model based on just two spectral vegetation indices, the green-red difference index and the blue-green difference index. We found that reasonable classification models could be generated using an RGB camera alone, and—more importantly—that one classification model could be used to detect coniferous seedling crowns across many sample plots in our study sites imaged under different lighting conditions. An example test-plot classification is shown in Figure 5.

3.2. Detection Accuracy, Error Patterns, and Density Estimates

To assess the detection accuracy in the test area, we considered a reference seedling detected (i.e., a true positive) if its corresponding geolocation point was inside a seedling object; otherwise, we considered the image object containing the point to be a false negative. Likewise, the seedling objects not containing a seedling geolocation point were considered false positives, and the rest of the non-seedling image objects were accordingly considered true negatives. The overall detection rate (sensitivity) for conifer seedlings in the independent test dataset was 75.8%: 113 out of the 149 seedlings surveyed in the test site were detected (Table 3). The classification model had a commission error (false positive) rate of 12.4% and an omission error (false negative) rate of 24.2%. The moderate imbalance between the omission and commission errors suggests that our workflow tends to underestimate the seedling density. This is understandable given the small size of the target seedlings and the complex environmental conditions in which they are found. The overall Kappa coefficient was 0.810 and the area under the receiver operating characteristic (ROC) curve was 0.93.

Table 2. Spectral, spatial, and textural attributes of image objects used for classification and regression tree (CART) modelling.

Metric Type	Description	Source
Spectral		
Brightness	Sum of all mean layer values within image object divided by number of layers	[27]
Border Contrast	Relative difference between brightness and mean intensity of image layers	[27]
BGVI	$(Mean\ Blue\ DN - Mean\ Green\ DN)$	[27]
EGI	$(2 * Mean\ Green\ DN) - Mean\ Red\ DN - Mean\ Blue\ DN$	[28]
GRDI	$(Mean\ Green\ DN - Mean\ Red\ DN)$	[29]
NBGVI	$\frac{(Mean\ Blue\ DN - Mean\ Green\ DN)}{(Mean\ Blue\ DN + Mean\ Green\ DN)}$	[30]
NDVI	$\frac{Mean\ NIRR\ DN - Mean\ Red\ DN}{Mean\ NIRR\ DN + Mean\ Red\ DN}$	[30]
NEGI	$\frac{(2 * Mean\ Green\ DN) - Mean\ Red\ DN - Mean\ Blue\ DN}{(2 * Mean\ Green\ DN) + Mean\ Red\ DN + Mean\ Blue\ DN}$	[31]
NGRDI	$\frac{(Mean\ Green\ DN - Mean\ Red\ DN)}{(Mean\ Green\ DN + Mean\ Red\ DN)}$	[28]
NGBDI	$\frac{(Mean\ Green\ DN - Mean\ Blue\ DN)}{(Mean\ Green\ DN + Mean\ Blue\ DN)}$	[28]
$R_{\bar{x}}, G_{\bar{x}}, B_{\bar{x}}, NIRR_{\bar{x}}, NIRG_{\bar{x}}, NIRB_{\bar{x}}$	$R_{\bar{x}} = \frac{\sum Red\ DN}{n}$, etc.	[27]
$R_{ratio}, G_{ratio}, B_{ratio}, NIRR_{ratio}, NIRG_{ratio}, NIRB_{ratio}$	$R_{ratio} = \frac{Mean\ Red\ DN}{Mean\ Red\ DN + Mean\ Green\ DN + Mean\ Blue\ DN}$, etc.	[27]
$R_{std}, G_{std}, B_{std}, NIRR_{std}, NIRG_{std}, NIRB_{std}$	Standard deviation of digital number (DN) values within image object	[27]
Spatial		
Border Index	Describes the ratio between the border length of an image object and the smallest enclosing rectangle	[27]
Asymmetry	Describes a length/width ratio between the image object and an approximated ellipse	[27]
Compactness (pixel)	The product of the length and width divided by the number of pixels	[27]
Compactness (polygon)	Ratio of the area of the image object to the area of a circle with the same perimeter	[27]
Perimeter	Pixel sum of the length of all edges in an image object	[27]
Pixel Area	Number of pixels contained in an image object	[27]
Roundness	Difference between the radius of the smallest enclosing and largest enclosed ellipse	[27]
Volume (voxel)	The number of volume elements (voxels) contained in an image object	[27]
Textural		
GLCM Contrast	Grey level co-occurrence matrix contrast	[27]
GLCM Homogeneity	Grey level co-occurrence matrix homogeneity	[27]

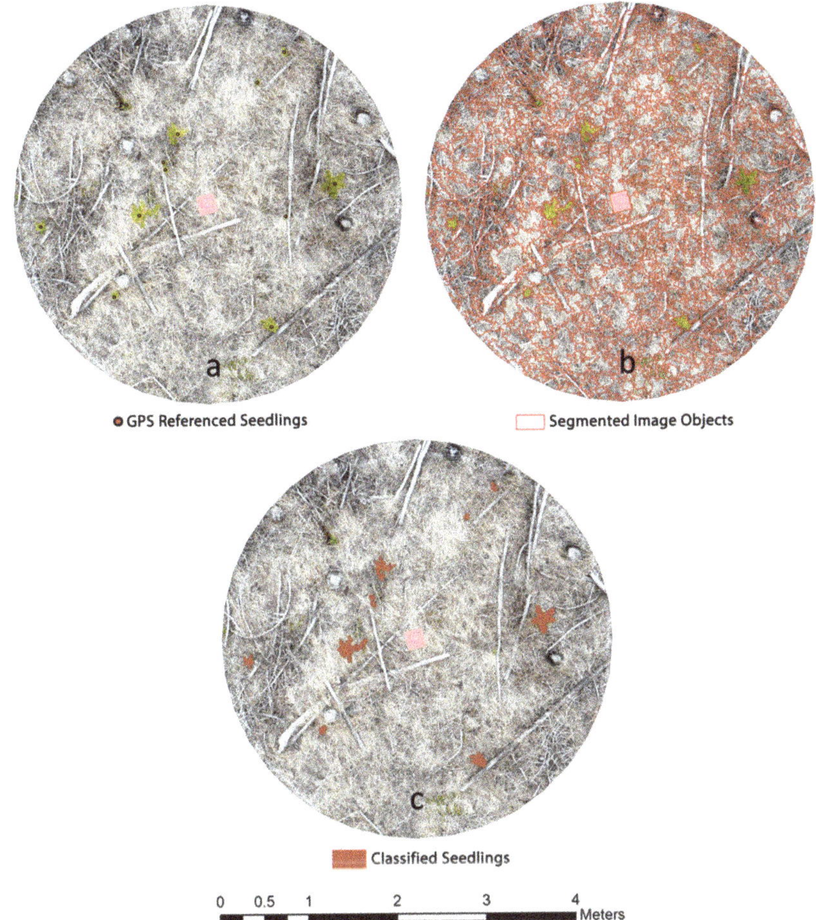

Figure 5. GNSS-surveyed reference seedlings (**a**) were surveyed in the field by trained personnel. UAV imagery was segmented into image objects (**b**) and subject to a decision-tree classification based on spectral indices from RGB imagery. The final output (**c**) delineates individual seedlings, whose accuracy was checked against the reference data.

We noted differential errors between the pine seedlings (86% detection rate, $n = 124$) and spruce seedlings (24% detection rate, $n = 25$), which were relatively rare in our test study area. This issue could likely be addressed using multiple classification models (one per species), although we did not attempt it here.

Plot-level associations between the CART-predicted stem numbers and field observations produced a Pearson's correlation coefficient of $r = 0.98$ in the test dataset (Figure 6). The plot accuracies in the test dataset ranged between 62.5% and 100%, with a mean accuracy of 86.2%. Once again, we observed a tendency of our workflow to underestimate when large numbers of seedlings are present (top end of Figure 6). This is not a critical problem, although, as plots with large numbers of small seedlings would be considered fully stocked, a moderate underestimation bias in these conditions can be tolerated in practical applications.

Table 3. Confusion matrix for the test data (independent validation) set. Interior matrix cells indicate both correctly classified objects (true positives [TP] and true negatives [TN]) and errors (false positives [FP] and false negatives [FN]). Square brackets in the interior cells break down the seedling reference data by species [pine/spruce].

		Reference	
		Seedling	Non-Seedling
CART	Seedling	113 (TP) [107/6]	16 (FP)
	Non-Seedling	36 (FN) [17/19]	8219 (TN)
		Sensitivity = 75.8%	Specificity = 99.7%

Seedling Commission Errors (false-positive rate) = 12.4%; Seedling Omission Errors; (false-negative rate) = 24.2%; Overall Accuracy = 99.4%; Kappa = 0.810.

Expressing the stem numbers derived from remote sensing on a per-hectare basis produced an estimated seedling density of 2160 stems/ha, versus 2500 stems/ha from the reference data. This represents an underestimation of 320 stems/ha (13.6%), which is consistent with the UAV data's tendency to slightly underestimate seedling counts, noted previously. Puliti et al. [20] estimated the stem numbers of mature trees in 38 fixed-area plots in Norway using photogrammetric data from UAVs, and reported root-mean-square errors of 538 stems/ha (39.2%). While less accurate than our results, the area-based regression analyses used by the authors is quite different than the object-detection approach used here.

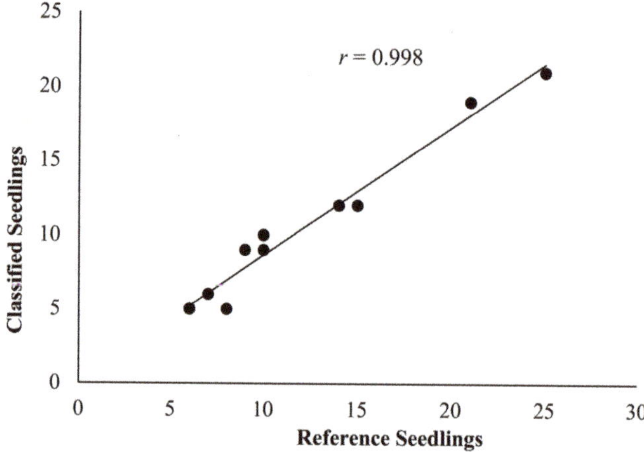

Figure 6. Association between classified seedling counts and field reference observations in the test-case study.

We could not find any published literature on the automated classification of very small seedlings (less than 5 years of age) against which to compare our results. Hall and Aldred [5] detected just 44% of the seedlings with a smaller than 30 cm crown diameter using the manual interpretation of 1:500 scale color-infrared imagery. Our results are slightly less accurate than those of Sperlich et al. [32], who reported an 88% overall accuracy from the photogrammetric detection of 219 mature tree crowns. Not surprisingly, our accuracy is lower than that in studies using UAVs to detect mature trees in plantation contexts, which present a simpler classification problem, in which individuals are spatially and structurally homogenous. For example, Torres-Sánchez et al. [21] reported a 95% accuracy using photogrammetric data on 54 olive trees, and Wallace et al. [17] reported a 98% detection rate of 308 eucalyptus plants in rows using UAV LiDAR data. Our results exceed those reported by Ke and Quackenbush [13] (70% user and producer accuracy) in their classification of individual trees in single-scene forest stands from piloted aerial imagery, and those of Chisolm et al. [33] (73% overall accuracy) in a below-canopy LiDAR survey of mature trees.

3.3. Challenges with Small Seedlings and Clusters of Seedlings

Many of the omission errors we encountered arose from image-segmentation challenges. Small individuals (down to 10 cm height with 5 cm crown diameter) and clusters of seedlings with contiguous crown types posed problems for our workflow (Figure 7). While millimetric spatial resolution helps identify fine features in our environment, it comes at the cost of spatial and spectral heterogeneity. While the CART algorithm is efficient at negotiating this heterogeneity, it can do so at the expense of specificity. To guard against this tendency, we pruned the decision tree to just two rules.

(a)   (b)   (c)

Figure 7. Despite using images with ultra-high spatial resolution, segmentation routines had difficulty detecting very small seedlings (**a**) or delineating groups of seedlings with contiguous crowns (**b**). Normally, individuals within clusters could only be delineated when small gaps occurred between crowns (**c**).

3.4. A Sample-Based Approach to Silvicultural Surveys

A significant portion of our study was devoted to working with a sample-based survey approach, rather than conventional wall-to-wall mapping: as is common in remote sensing. This approach achieves significant time savings in terms of field data collection and avoids additional photogrammetric and orthomosaic processing costs. We estimate that a standard wall-to-wall UAV survey over the training site would require a flight time of 33 minutes, during which the platform would fly 9.8 km (Figure 8a). Alternatively, a sample-based survey of the same area could take place in under six minutes and cover just 2.8 km, with a time savings of 81% (Figure 8b). While we acknowledge that wall-to-wall surveys of forest-harvest areas the size of the ones we worked in are currently possible with UAV platforms, and that wall-to-wall surveys (i.e. census) may provide incremental benefits to sampling, we contend that sample-based flight planning is currently underutilized by the UAV community, and may be crucial to developing operational workflows.

Our workflow is based on spectral variables from a standard RGB camera alone, with no geometric pre-processing and no secondary photogrammetric products. This approach has its pros and cons. For example, the exclusion of spatial and structural variables from our workflow limits our approach to applications where seedlings are spectrally distinct from their surroundings; hence, our requirement for seasonal leaf-off conditions. The main benefit is a streamlined workflow that simplifies the survey and processing procedures. Nex and Remondino [34] estimate that the placement of ground control points and photogrammetric processing constitute 55% of the time effort required to perform a photogrammetric UAV survey, compared to 25% for flight planning and image acquisition. Alternative workflows to ours that incorporate multiple datasets often require precise geometric integration using ground-control points, which must be laid out and surveyed prior to image acquisition. Photogrammetric processing also adds significant computational costs, and may introduce ground-object distortions [35], moaicking artefacts [36], and radiometric inconsistencies. For example, Borgogno-Mondino et al. [37] explain how color-balancing algorithms embedded in commercial image-processing packages can degrade the radiometric quality of the resulting orthomosaics and limit the effectiveness of derived spectral indices. These issues will certainly diminish in time, given the rapid advancement in direct georeferencing technology [34,38], alternative spatial processing routines [39], and integrated sensor systems. As a result, we expect future studies to find incremental value in photogrammetric data for forest-regeneration surveys, as other authors have reported with the detection of mature trees [2,21,32]. In the meantime, the benefits of the simplified workflows for operational projects are substantial.

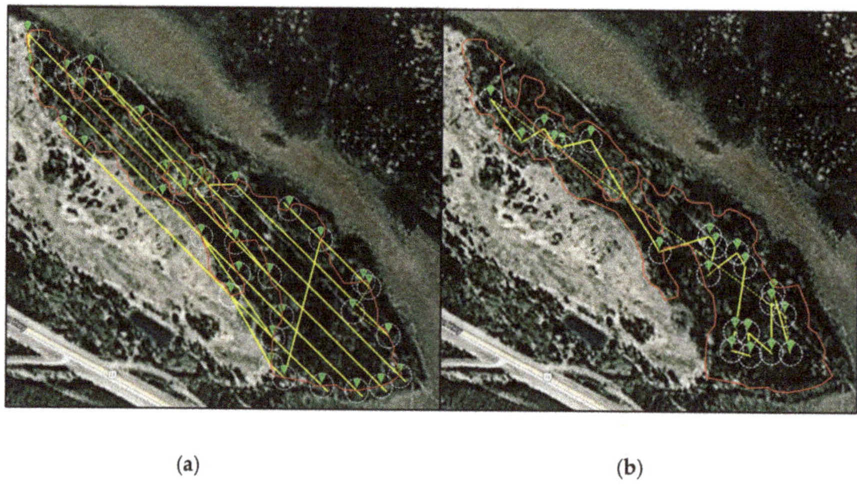

(a)   (b)

Figure 8. The flight plan for a standard wall-to-wall survey (**a**) is almost six times longer in duration and three times longer in distance than the sample-based approach (**b**) used in this research.

Despite these benefits, our simplified sample-based workflow also contains a number of drawbacks. The lack of geometric correction means that images contain relief distortion and optical lens distortion, which introduce variability into the seedlings' appearance. Excluding the geometric correction from the workflow also introduced variability into the spatial resolution of the resulting images, with the ground-sample distance (GSD) being just a simple function of the flying altitude and camera lens focal length. We reduced this scale effect by outfitting our UAV with a laser rangefinder, which allowed us to hover at exactly 15 m above ground level during imaging. With this, we ensured a common GSD of 3 mm at nadir (5 mm for the NIR camera), which became 3.2 mm at the edge of the circular plot over flat terrain. Even in slightly uneven ground (slopes in our study area did not exceed 10%), the GSD at the low side of a sloping lot did not exceed 3.4 mm. A standard UAV equipped with a conventional GNSS and barometer would be unable to acquire images in such a consistent manner. While the seedlings at the edge of plots appeared up to 30% smaller than those at the center, this effect is unlikely to decrease the detection rate, except perhaps for very small seedlings arising from natural regeneration. The occlusion by taller vegetation could also impair the detection of those seedlings close to the edge of the plots. Once again, though, this is unlikely to decrease the detection, because most shrubs were devoid of leaves at the acquisition date, and there were no saplings or mature conifer trees within the plots. Additional challenges in our workflow arose from the substantial spectral variability caused by the changing illumination conditions during our flights. While this variability could be reduced with the use of an integrated irradiance sensor, operational workflows could benefit from several classification models designed to account for diversity in seedling species, radiometric conditions, and surrounding vegetation, as well as for other sources of spectral variability. We encourage future researchers to assess the value of incorporating scene-level variables that categorize samples on the basis of brightness, time of day, latitude, greenness, and other factors.

Finally, we note that the Regeneration Standards of Alberta [1] call for between 2.77 and 12.4 sample plots per hectare, depending the size of the harvest area being assessed. In a real-world scenario, this means that we would need to increase our sampling intensity substantially (4 or 5 fold) over the design used in this research. However, it was not our intention to conduct actual stocking assessments, but rather to create and evaluate a workflow that could perform efficiently in this respect.

4. Conclusions

In this study, we assessed the capacity of optical photography from an unmanned aerial vehicle (UAV), to perform coniferous-seedling detection in an object-based environment. The 75.8% overall detection rate of the ground-surveyed seedlings in an independent test site ($n = 149$) demonstrates the utility of our approach. Error analytics revealed a slight tendency to underestimate seedlings, although the plot-level associations with ground surveys were very high ($r = 0.98$, $n = 14$). Red-edge imagery offered no significant advantage over the data from a standard RGB camera, and our final decision tree was comprised of a simple two-rule model based on just two spectral vegetation indices, the Green-Red Difference Index and the Blue-Green Difference Index. We found spatial and textural variables to be unnecessary for identifying coniferous seedlings in the conditions we assessed. Our workflow relies on seasonal leaf-off conditions when grasses are senesced and seedlings are spectrally distinct from their surroundings, and it would not be expected to perform with the same efficiency at other times of the year, or with deciduous seedlings.

The seedling surveys conducted with UAVs are feasible and efficient, but further research is required. For example, we expect that the increased variability encountered under operational conditions will require the complementary use of several models or the application of more sophisticated machine-learning approaches. We encourage other researchers to explore the detectability of other seedling species in new environments, and at different times of the year. Also, a full stocking-assessment workflow would require delineated seedlings to be assessed for other attributes, including height, species, and condition: challenges that will probably require enhanced spectral information and 3D data collected using light detection, ranging or photogrammetry.

The trend towards quantified vegetation surveys at this level of detail is a promising development, both for forest management and in the larger context of restoration. Perhaps the most useful area for the future development of the use of UAVs in forest management is repeatability. This is a progressive approach, as the future use of UAVs may not depend as much on the correlation of the metrics derived from the data collected with unmanned aerial vehicles to field observations as on simple data and mensuration consistency. Wallace et al. [18] have been pioneers in this regard, publishing a study focused on the repeatability of the measurements taken using unmanned aerial vehicles. We encourage other researchers to assist in the development and reporting of forest mensuration workflows based on remote sensing to establish a body of literature that provides a foundation from which the consistency and repeatability of these novel techniques can be assessed.

Author Contributions: C.F. conceived of the study, performed the field work, operated the UAV, completed the data analysis, and wrote the first draft of the manuscript. C.F., G.J.M., and G.C. designed the experiments and interpreted the results. G.C. and G.J.M. edited the manuscript and prepared the final draft for publication.

Acknowledgments: This research is part of the Boreal Ecosystem Recovery and Assessment (BERA) project (www.bera-project.org), and was supported by a Natural Sciences and Engineering Research Council of Canada Collaborative Research and Development, Grant (CRDPJ 469943-14), in conjunction with Alberta-Pacific Forest Industries, Cenovus Energy, ConocoPhillips Canada, and Devon Canada Corporation. The BERA funding facilitated both the research activities and subsequent open-access publishing of this work. Jamie Miller (Weyerhaeuser Grand Prairie) and Victor Fobert (Weyerhaeuser Pembina) provided access to the forest-harvest blocks used for this study. We also thank Jennifer Thomas for her editorial review. The comments of three anonymous reviewers helped us improve the manuscript.

Conflicts of Interest: The authors declare no conflict of interest. The funding sponsors had no role in the design of the study; in the collection, analyses, or interpretation of data; in the writing of the manuscript; or in the decision to publish the results.

References

1. Alberta Environment and Sustainable Resource Development (AESRD). *Regeneration Standards of Alberta*; Department of Environment and Sustainable Resource Development: Edmonton, AB, Canada, 2015.
2. Puliti, S.; Talbot, B.; Astrup, R. Tree-stump detection, segmentation, classification, and measurement using unmanned aerial vehicle (UAV) imagery. *Forests* **2018**, *9*, 120. [CrossRef]
3. Peña, J.M.; Torres-Sanchez, J.; Serrano-Perez, A.; de Castro, A.I.; Lopez Granados, F. Quantifying efficacy and limits of unmanned aerial vehicle (UAV) technology for weed seedling detection as affected by sensor resolution. *Sensors* **2015**, *15*, 5609–5626. [CrossRef] [PubMed]
4. Kirby, C.L. *A Camera and Interpretation System for Assessment of Forest Regeneration*; Environment Canada, Canadian Forestry Service, Northern Forest Research Centre: Edmonton, AB, Canada, 1980.
5. Hall, R.J.; Aldred, A.H. Forest regeneration appraisal with large-scale aerial photographs. *For. Chron.* **1992**, *68*, 142–150. [CrossRef]
6. Goodbody, T.R.; Coops, N.C.; Hermosilla, T.; Tompalski, P.; Crawford, P. Assessing the status of forest regeneration using digital aerial photogrammetry and unmanned aerial systems. *Int. J. Remote Sens.* **2017**, *38*, 1–19. [CrossRef]
7. Agarwal, S.; Vailshery, L.S.; Jaganmohan, M.; Nagendra, H. Mapping urban tree species using very high resolution satellite imagery: Comparing pixel-based and object-based approaches. *ISPRS Int. J. Geo-Inf.* **2013**, *2*, 220–236. [CrossRef]
8. Gougeon, F.A. A crown-following approach to the automatic delineation of individual tree crowns in high spatial resolution aerial images. *Can. J. Remote Sens.* **1995**, *21*, 274–284. [CrossRef]
9. Wulder, M.; Niemann, K.O.; Goodenough, D.G. Local maximum filtering for the extraction of tree locations and basal area from high spatial resolution imagery. *Remote Sens. Environ.* **2000**, *73*, 103–114. [CrossRef]
10. Franklin, S.E.; Hall, R.J.; Smith, L.; Gerylo, G.R. Discrimination of conifer height, age and crown closure classes using Landsat-5 TM imagery in the Canadian Northwest Territories. *Int. J. Remote Sens.* **2003**, *24*, 1823–1834. [CrossRef]
11. Pouliot, D.A.; King, D.J.; Pitt, D.G. Development and evaluation of an automated tree detection-delineation algorithm for monitoring regenerating coniferous forests. *Can. J. For. Res.* **2005**, *35*, 2332–2345. [CrossRef]
12. Wolf, B.M.; Heipke, C. Automatic extraction and delineation of single trees from remote sensing data. *Mach. Vision Appl.* **2007**, *18*, 317–330. [CrossRef]
13. Ke, Y.H.; Quackenbush, L.J. A comparison of three methods for automatic tree crown detection and delineation from high spatial resolution imagery. *Int. J. Remote Sens.* **2011**, *32*, 3625–3647. [CrossRef]
14. Leckie, D.G.; Gougeon, F.; McQueen, R.; Oddleifson, K.; Hughes, N.; Walsworth, N.; Gray, S. Production of a large-area individual tree species map for forest inventory in a complex forest setting and lessons learned. *Can. J. Remote Sens.* **2017**, *43*, 140–167. [CrossRef]
15. Panagiotidis, D.; Abdollahnejad, A.; Surovy, P.; Chiteculo, V. Determining tree height and crown diameter from high-resolution UAV imagery. *Int. J. Remote Sens.* **2017**, *38*, 2392–2410. [CrossRef]
16. Fritz, A.; Kattenborn, T.; Koch, B. UAV-based photogrammetric point clouds—Tree stem mapping in open stands in comparison to terrestrial laser scanner point clouds. In Proceedings of the International Archives of the Photogrammetry, Remote Sensing and Spatial Information Sciences, UAV-g2013, Rostock, Germany, 4–6 September 2013; pp. 141–146.
17. Wallace, L.; Lucieer, A.; Watson, C.S. Evaluating tree detection and segmentation routines on very high resolution UAV LiDAR. *IEEE Trans. Geosci. Remote Sens.* **2014**, *52*, 7169–7628. [CrossRef]

18. Wallace, L.; Musk, R.; Lucieer, A. An assessment of the repeatability of automatic forest inventory metrics derived from UAV-borne laser scanning data. *IEEE Trans. Geosci. Remote Sens.* **2014**, *52*, 7160–7169. [CrossRef]

19. Jaskierniak, D.; Kuczera, G.; Benyon, R.; Wallace, L. Using tree detection algorithms to predict stand sapwood area, basal area and stocking density in *Eucalyptus regnans* forest. *Remote Sens.* **2015**, *7*, 7298–7323. [CrossRef]

20. Puliti, S.; Orka, H.O.; Gobakken, T.; Naesset, E. Inventory of small forest areas using an unmanned aerial system. *Remote Sens.* **2015**, *7*, 9632–9654. [CrossRef]

21. Torres-Sanchez, J.; Lopez-Granados, F.; Serrano, N.; Arquero, O.; Pena, J.M. High-throughput 3-D monitoring of agricultural-tree plantations with unmanned aerial vehicle (UAV) technology. *PLoS ONE* **2015**, *10*, e0130479. [CrossRef] [PubMed]

22. Kang, J.; Wang, L.; Chen, F.; Niu, Z. Identifying tree crown areas in undulating eucalyptus plantations using JSEG multi-scale segmentation and unmanned aerial vehicle near-infrared imagery. *Int. J. Remote Sens.* **2017**, *38*, 2296–2312. [CrossRef]

23. Nevalainen, O.; Honkavaara, E.; Tuominen, S.; Viljanen, N.; Hakala, T.; Yu, X.W.; Hyyppa, J.; Saari, H.; Polonen, I.; Imai, N.N.; et al. Individual tree detection and classification with UAV-based photogrammetric point clouds and hyperspectral imaging. *Remote Sens.* **2017**, *9*, 185. [CrossRef]

24. Hirschmugl, M.; Ofner, M.; Raggam, J.; Schardt, M. Single tree detection in very high resolution remote sensing data. *Remote Sens. Environ.* **2007**, *110*, 533–544. [CrossRef]

25. Chen, S.J.; McDermid, G.J.; Castilla, G.; Linke, J. Measuring vegetation height in linear disturbances in the boreal forest with UAV photogrammetry. *Remote Sens.* **2017**, *9*, 1257. [CrossRef]

26. Hird, J.N.; Montaghi, A.; McDermid, G.J.; Kariyeva, J.; Moorman, B.J.; Nielsen, S.E.; McIntosh, A.C.S. Use of unmanned aerial vehicles for monitoring recovery of forest vegetation on petroleum well sites. *Remote Sens.* **2017**, *9*, 413. [CrossRef]

27. *Trimble eCognition Developer 8.8 Reference Book*, Trimble Germany GmbH: Munich, Germany, 2015.

28. Woebbecke, D.M.; Meyer, G.E.; Vonbargen, K.; Mortensen, D.A. Color indexes for weed identification under various soil, residue, and lighting conditions. *Trans. ASAE* **1995**, *38*, 259–269. [CrossRef]

29. Loris, V.; Damiano, G. Mapping the green herbage ratio of grasslands using both aerial and satellite-derived spectral reflectance. *Agric. Ecosyst. Environ.* **2006**, *115*, 141–149. [CrossRef]

30. Jannoura, R.; Brinkmann, K.; Uteau, D.; Bruns, C.; Joergensen, R.G. Monitoring of crop biomass using true colour aerial photographs taken from a remote controlled hexacopter. *Biosys. Eng.* **2015**, *2015*. *129*, 341–351. [CrossRef]

31. Jensen, J.R. *Remote Sensing of the Environment: An Earth Resource Perspective*; Prentice Hall: Upper Saddle River, NJ, USA, 2007.

32. Sperlich, M.; Kattenborn, T.; Koch, B. Potential of unmanned aerial vehicle based photogrammetric point clouds for automatic single tree detection. *Gemeinsame Tagung* **2015**, 1–6.

33. Chisholm, R.A.; Cui, J.; Lum, S.K.Y.; Chen, B.M. UAV LiDAR for below-canopy forest surveys. *J. Unmanned Veh. Syst.* **2013**, *1*, 61–68. [CrossRef]

34. Nex, F.; Remondino, F. UAV for 3D mapping applications: A review. *Appl. Geomatics* **2014**, *6*, 1–15. [CrossRef]

35. Rosnell, T.; Honkavaara, E. Point cloud generation from aerial image data acquired by a quadrocopter type micro unmanned aerial vehicle and a digital still camera. *Sensors* **2012**, *12*, 453–480. [CrossRef] [PubMed]

36. Lin, C.H.; Chen, B.H.; Lin, B.Y.; Chou, H.S. Blending zone determination for aerial orthimage mosaicking. *ISPRS J. Photogramm. Remote Sens.* **2016**, *119*, 426–436. [CrossRef]

37. Borgogno-Mondino, E.; Lessio, A.; Tarricone, L.; Novello, V.; de Palma, L. A comparison between multispectral aerial and satellite imagery in precision viticulture. *Prec. Agric.* **2018**, *19*, 195–217. [CrossRef]

38. Klingbeil, L.; Eling, C.; Heinz, E.; Wieland, M.; Kuhlmann, H. Direct georeferencing for portable mapping systems: In the air and on the ground. *J. Surv. Eng.* **2017**, *143*, 04017010. [CrossRef]

39. Ribeiro-Gomes, K.; Hernandez-Lopez, D.; Ballesteros, R.; Moreno, M.A. Approximate georeferencing and automatic blurred image detection to reduce the costs of uav use in environmental and agricultural applications. *Biosyst. Eng.* **2016**, *151*, 308–327. [CrossRef]

MDPI
St. Alban-Anlage 66
4052 Basel
Switzerland
Tel. +41 61 683 77 34
Fax +41 61 302 89 18
www.mdpi.com

Forests Editorial Office
E-mail: forests@mdpi.com
www.mdpi.com/journal/forests